MAKING OF THE MODERN WORLD

Driving wheel
of the steam locomotive *Rocket*
(*See* STEPHENSON'S 'ROCKET')

MAKING OF THE MODERN WORLD

MILESTONES OF SCIENCE AND TECHNOLOGY

EDITED BY

NEIL COSSONS

WITH ANDREW NAHUM AND PETER TURVEY

PHOTOGRAPHS BY

PHILIP SAYER

JOHN MURRAY

IN ASSOCIATION WITH THE

SCIENCE MUSEUM

Text and special photography © Science Museum

First published in 1992
by John Murray (Publishers) Ltd
50 Albemarle Street
LONDON W1X 4BD

A catalogue record for this book
is available from the British Library

ISBN 0-7195-5121-8

Typeset in 8/12pt Linotron Joanna
by Wearset, Boldon, Tyne and Wear

Printed and bound in Italy
by Graphicom, Vicenza

Designed by Ken Garland and Associates

CONTENTS

The Science Museum inventory numbers appear
against the illustrations

INTRODUCTION

NEIL COSSONS
DIRECTOR OF THE SCIENCE MUSEUM

The collections of the Science Museum – the National Museum of Science & Industry – provide an extraordinary and vivid insight into the progress of mankind in the last 200 to 300 years. That these collections are overwhelmingly of British origin is a reflection of the country's rise as the first industrial nation in the middle years of the eighteenth century and of its role as 'workshop of the world' in the nineteenth. But their significance transcends the national context: they celebrate the emergence of modern scientific and industrial civilization.

The circumstances that led to the creation of the Science Museum and to the immense diversity and richness of its collections are twofold: on the one hand there was concern to improve scientific and technical education in the middle years of the nineteenth century, on the other was a desire to record and celebrate the great achievements of scientists and engineers. This tension, between the need to educate and instruct and the wish to preserve and record, forms a continuous thread throughout the history of the Science Museum and of its great sister institutions overseas.

Thus in 1911 when the committee chaired by Sir Hugh Bell Bt (1844–1934) to consider the future of the Science Museum issued the first part of its report, it recognized this dual purpose. Under a section entitled 'Purposes the Science Museum should serve' the report states that 'So far as is possible by means of exhibited scientific instruments and apparatus, machines and other objects, the Collections in the Science Museum ought to afford illustrations and exposition of the various branches of Science within its field and of their applications in the Arts and Industries. The Museum ought also to be a worthy and suitable house for the preservation of appliances which hold honoured place in the progress of Science or in the history of invention'.

Reconciling the conservation of collections with interpretation and education, an issue common to all museums, raises special problems in museums of science and industry where the objects do not al-

HRH Prince Albert (1819–61)

ways explain themselves. In the case, for example, of products of the solid-state age – such as a microprocessor – form offers little insight into function.

But for the purposes of scholarly understanding, and as a research record to be interrogated at will, the collections require no justification. They are there as evidence. Equally, to present them as part of our culture is becoming daily more acceptable. We are at last recognizing that science and technology are distinctive characteristics of our success as a society and that we need to value those who engage in science and technology not only for what they do but for how they do it. The collections of the Science Museum are evidence both of the content of science, of engineering, industry and medicine, and also of the process. They are also a human record displaying the frailties, failings and limitations of progress as much as the glories of its achievement.

Museums are about collections. It is this that distinguishes them from all other places of learning and scholarship, research, education, enlightenment and entertainment. And it is collections that separate science museums from science centres and exploratoria. Whilst the former may engage in some of the activities of the latter, the opposite cannot be so. And yet such is the nature of our dichotomous society that we have difficulty in recognizing the importance of the material record of scientific and technological history. Rarely do these collections enjoy the romantic aura of antiquity; it is often fortuitous that they offer any aesthetic pleasure.

But when we view the great collections of the National Museum of Science & Industry and gain some understanding of the extraordinary events they reflect we can begin to appreciate why the world today is the way it is. The remarkable scale and stature of the Museum becomes apparent when we add to that a knowledge of its history and of the prizes it holds.

THE ORIGINS OF THE SCIENCE MUSEUM

If any one man could be said to be the father of the Science Museum then it is His Royal Highness Prince Albert (1819–61), Consort of Queen Victoria. He was a leading light in the movement to improve scientific and technical education. He was also the driving force behind the Great Exhibition of 1851. He proposed that its profits, of £186,436, be used to purchase land south of Kensington Gore and to establish there institutions devoted 'to the furtherance of industrial pursuits of all nations'. In a memorandum dated 20 August 1853, he set out 'a general plan for the buildings it is proposed to erect on the newly purchased ground at South Kensington', including what were described as 'Museums or Schools of Science and Industry'. The other immediate outcome of this movement, also urged by the Prince, was the setting up by the Government of the Science and Art Department in March 1853. Lyon Playfair

'Sketch taken from Whitworth's stand
of machine tools, for planing, slotting, drilling, boring etc. . .'
at the Great Exhibition, 1851

(1818–98), chemist and scientific administrator, was appointed Science Secretary. He was among those who recognized that Britain's world leadership in industry was due both to her natural resources and her head start, but that she would 'recede as an industrial nation, unless her industrial population became much more conversant with science than they are now'. The aim of the Science and Art Department was to 'increase the means of industrial education and to extend the influence of science and art on productive industry'. Its plans included 'museums by which all classes might be induced to investigate those common principles of taste which may be traced in the works of excellence of all ages'.

Those objects from the Great Exhibition that had been acquired for preservation were brought together at Marlborough House in 1852 to form, with others from the School of Design, the beginnings of a museum of ornamental art. In 1856 the collections were removed to South Kensington and housed in a corrugated-iron building, soon nicknamed the Brompton Boilers. There was a preponderance of art objects but the rest (more accurately described as non-art objects than the Science Collections by which they were generally known) were many and varied. They included models of machinery and industrial plant, collections illustrating foods and animal production, examples of structures and educational materials – books, models and apparatus for use in primary education. On 24 June 1857 the South Kensington Museum was opened to the public for the first time.

In 1864 the Royal School of Naval Architecture and Marine Engineering was established at South Kensington by the Admiralty and with it was inaugurated in the South Kensington Museum a shipping and maritime engineering collection. Although the collection consisted at first mainly of models lent by the Admiralty, it was further developed in the ensuing years by the acquisition of models obtained from engineering, shipbuilding and ship owning firms, and from Lloyds Register of British and Foreign

Bennet Woodcroft (1803–79),
founder of the Patent Office Museum

Shipping. In 1873 the School left South Kensington and was transferred to Greenwich together with the Admiralty's own models. The rest remained to form the core of what was to become an outstanding collection of shipping, naval architecture and marine engineering.

In 1874 the Royal Commission on Scientific Instruction and the Advancement of Science produced its final report. It recommended the formation of a loan collection of scientific apparatus and an exhibition. A powerful committee headed by the Lord Chancellor and including the presidents of the professional institutions and learned societies was formed to advise. The collection was to include 'not only apparatus for teaching and investigation, but also such as possessed historic interest on account of the persons by whom, or the researches in which, it had been employed'. And the exhibition, which was opened in May 1876 by Queen Victoria, was to be

international in scope; several European countries as well as the United States accepted the invitation to participate and the organizing committee was augmented by their representatives.

In the early 1880s the curators of the science collections and the professors and students of the science colleges that had by then transferred to South Kensington were finding library provision inadequate. The Government remedied this in 1883 by setting up the Science Library (now the Science Museum Library). It was formed by merging the science books from the South Kensington Museum's science and educational collections and the library of the Museum of Practical Geology in Jermyn Street, developed from the original gift of Sir Henry de la Beche (1796–1855).

BENNET WOODCROFT AND THE PATENT MUSEUM

Then, in 1884 came a major influx to the science collections, namely the Patent Museum, a collection of 'patent models' consisting both of replicas for instructional purposes and, more importantly, actual specimens of celebrated machines, built up by Bennet Woodcroft (1803–79), Assistant to the Commissioners of Patents from 1852 and founder of the Patent Office Library.

Bennet Woodcroft was born at Heaton Norris near Stockport. He became apprenticed to a silk weaver and spent his first forty or so years in the North of England apparently devoting much of his time to invention in fields as diverse as textile machinery and marine propulsion. In 1845 a screw propeller to a design by Woodcroft was fitted to I K Brunel's new iron steamship *Great Britain* which had just completed its first return passage to the United States. At about the same time Woodcroft was working on a variable-pitch propeller and later, in the 1850s, he was one of those who shared the £20,000 award voted by Parliament to a group of engineers whose development of the screw had benefited the Royal Navy. Woodcroft was clearly well regarded by his engineer contemporaries, notably Fairbairn, Eaton Hodgkinson, Whitworth and Nasmyth, and in 1847 he applied successfully for the professorship of the Mechanical Principles of Engineering at University College, London. In recommending him John Graham, the Manchester chemist, wrote, 'He is extensively acquainted with the *history* and uses of machines'. It was not a statement that could have been made of many engineers then or now.

Woodcroft's position at University College was shortlived. His background in invention led him to a position on the Society of Arts committee on patent reform, then at a crucial stage in its work, and in 1852 he was appointed Assistant to the Commissioners of Patents and effectively head of the first office for handling the technical side of patents to be established in England.

In this capacity Woodcroft became involved in the debate, after the 1851 Exhibition, on the setting up of a museum. When he arrived at the Patent Office in 1852 he already had a large collection of 'models' – later reported as numbering some 900 – and in 1856 they were described as forming the nucleus of a 'National Collection of Models of Invention'. This was the Patent Office Museum and it found a home in buildings adjoining the newly opened South Kensington Museum.

Woodcroft was a relentless but selective collector. At first the idea of a patent museum was linked with the models submitted as part of the process of patenting, as was the established practice in the United States. But it is clear that it was the historically significant invention and the educational demonstration of principles that were his prime concerns. In this farsightedness Bennet Woodcroft occupies a position of profound importance in the origins of the Science Museum's collections.

It was at his insistence that both *Puffing Billy* and Stephenson's *Rocket* were preserved. He pursued and successfully captured Symington's marine engine and, in 1864, arranged for the contents of James Watt's home workshop to be collected, his idea

A label from the Patent Museum;
Richard Trevithick's high-pressure boiler

being 'to build a room which should be the very "counterfeit presentment" of the classic garret, and to replace every article in the precise position which it occupied at Heathfield'. Woodcroft also collected portraits under the title of 'The National Gallery of Portraits of Inventors'. Finally, in 1884, some five years after Bennet Woodcroft's death, the collection of the Patent Office Museum passed to the South Kensington Museum. To quote the official history, 'the Museum might then be said to have begun to assume the form of a national museum of science and industry'. In collecting celebrated machines, and thus immortalizing their inventors, Woodcroft made a major contribution to the development of museums and although the concept of *in situ* preservation is more recent, his assiduity in the recovery of classics of industrial development entitles him to be called one of the founders of industrial archaeology.

GROWTH AND DEVELOPMENT

The arts and science collections, although still together in the same building, had gradually been assuming separate and distinct identities and this was recognized in 1885 by naming the latter the Science Museum. In 1899 Queen Victoria laid the foundation

stone for the new Art Museum on the east side of Exhibition Road. She ordained that the new museum should be named the Victoria & Albert Museum. For a while this name was applied to the South Kensington Museum as a whole, embracing both art and science. However, with the opening of the new buildings by King Edward VII in 1909 came a formal separation of the Science Museum and the Art Museum, the latter taking the title of the Victoria & Albert Museum all to itself.

By now the site for a building to house the science collections had been identified on the west side of Exhibition Road between the Natural History Museum and the Royal College of Science. Having fought off a threat to build a new art gallery there (this was eventually to become the Tate Gallery) and with the Victoria & Albert Museum safely in its new buildings, it was time to turn to the proper housing of the Science Museum.

The government appointed a committee under the chairmanship of Sir Hugh Bell, prominent in the iron and steel industry, to consider the aims of the Museum and how they might best be achieved. The Bell Committee completed its report in 1912 and its recommendations were accepted. Work started on the East Block of a new building the following year. Indeed, the Bell Report is the foundation of the Science Museum and its development up to the present day. Its recommendations, which still stand as guiding principles for the Museum and its development, have only been partly implemented so that even now, despite the considerable scale and diversity of the Museum and its activities, and the huge growth in its collections, only two-thirds of what was planned eighty years ago has yet been built.

With the shell of the new building on Exhibition Road complete, the First World War prevented it being taken up for Museum purposes. But in 1920 Sir Henry Lyons (1864–1944) took over as Director and it was under thirteen years of his inspired leadership that the Science Museum rose to world stature. The new Museum was formally opened by His Majesty

King George V in 1928; by then a number of other recommendations of the Bell Committee had already been implemented. Temporary exhibitions became a feature of the Museum's programme, notably one in 1919 marking the centenary of the death of James Watt, and in 1924 a guide lecturer service was inaugurated. In 1931 the Children's Gallery was opened. This was to become one of the most popular features of the Museum and helped to raise the number of visitors to an estimated million and a quarter annually by 1933, more than any other national museum.

The Science Museum closed shortly after the outbreak of the Second World War; the objects were evacuated to places of safety and many of the staff transferred to war service. Only the Library remained open, satisfying wartime demand for scientific literature. With the coming of peace the Museum opened again in February 1946. It was at last decided to erect the second stage of the building envisaged by Bell and by 1951 the Centre Block was partially complete, just in time to accommodate the Science Section of the Festival of Britain. It was not finally completed however until 1961. Then began a period of twenty years of steady expansion with further space being added in the 1970s. In 1980–1 the objects from the Wellcome Museum of the History of Medicine, which the Wellcome Trust had placed with the Museum in 1976, were put on display. This outstanding collection relating to medical history, a subject new to the Museum, was one of the most important acquisitions in its history and significantly broadened its portfolio of activities which today span the physical and medical sciences, as well as engineering and industry.

The Library too was given a new direction during the 1970s. It had lost its national lending function in 1962 and during the 1960s its neighbour Imperial College had considerably developed its own library services. In 1971 the Government directive that led to the establishment of the British Library indicated that the Science Museum Library should be de-veloped as a national reference library of the history of science and technology, with particular emphasis on supporting the work of the curators in the other departments of the Museum. The Science Museum Library moved into new accommodation contiguous with the Lyon Playfair Library of Imperial College and there is now a close working relationship between the two.

To date the Museum's public activities had been concentrated within its building in South Kensington although substantial reserve collections were housed in warehouses elsewhere in London. In the early 1970s, however, another significant development took place. Under the terms of the Transport Act 1968 the collections of railway material that had been built up under the auspices of the British Transport Commission since 1951 were transferred to the Department of Education & Science, of which the Science Museum was by then a part. It was decided that a new national museum should be developed to house this outstanding railway material and after prolonged debate about its location it was decided to convert the North Motive Power Depot at York, which had recently closed as a result of the change-over from steam to diesel-electric traction. York was a good choice for a new National Railway Museum. A railway centre in its own right, with a long and distinguished history, it was also the home to Britain's first railway museum which had been opened in 1928 by the London & North Eastern Railway in small premises on Queen Street. With a decision to establish the National Railway Museum at York, under the direction of the Science Museum, the National Railway Collection had a new and permanent home. The Queen Street museum was closed and in September 1975, the National Railway Museum in York was opened by His Royal Highness the Duke of Edinburgh.

In 1983, under the terms of the National Heritage Act, the Museum was transferred from a Government department to a Board of Trustees appointed by the Prime Minister, thus bringing it into line with the other major national museums and ending a tradition going back 127 years. Today the Museum is funded by grant-in-aid from the Department of National Heritage, restoring at least symbolically the link between arts and manufactures envisaged by Prince Albert.

Further opportunities for development outside London arose during the 1980s. The British prototype of the Anglo-French supersonic airliner Concorde 002 had been taken into the Science Museum's National Aeronautical Collection and a home had to be found for it. In 1980 it was moved into a purpose-built museum building adjoining the Fleet Air Arm Museum at Yeovilton in Somerset.

On a much larger scale was the establishment of the National Museum of Photography, Film & Television in Bradford, the first part of which opened in 1983. This Museum now houses most of the Museum's photographic material including the outstanding collection of Talbot images, transferred from South Kensington in 1989. At the heart of this third major component of the National Museum of Science & Industry is a large-screen IMAX cinema; a second screen and auditorium opened in 1992. There are galleries on the history of television and the Kodak Museum charts the development of popular photography. Today the National Museum of Science & Industry serves an audience of some 3,000,000 people each year at its major locations in London and Yorkshire.

THE INTERNATIONAL CONTEXT

There is no direct equivalent overseas of the National Museum of Science & Industry. Certainly no other museum can match the extraordinary depth and diversity of its collections. But in looking at the way in which the world's leading museums of science, engineering and industry have developed – and there are less than half a dozen that attempt to span the whole field – a number of common themes emerge. The emphasis on progress and on achievement, both national and personal, is universal. Bennet Woodcroft's enthusiasm for portraits of engineers and examples of their inventions was influenced by the collections of the Conservatoire National des Arts et Métiers in Paris, inaugurated in 1794, and by the highly organized United States Patent Office, in which a model had to be deposited as part of each patent application. The significance of national and international exhibitions and later even of the Olympic Games movement – in the setting up of the Tekniska Museet in Stockholm, for example – has also been considerable. So too have been the difficulties in establishing priorities for funding museums of science and industry in the face of the competing needs of other museums.

Oskar von Miller (1855–1934), the guiding genius and driving force behind the Deutsches Museum in Munich, trained as a civil engineer and entered government service in 1878. In 1881 he was sent to an international exhibition of electrical engineering in Paris and later he organized trade fairs in Munich and Frankfurt. He also visited the Conservatoire and the South Kensington Museum. The Museum he founded in 1903 was called the Deutsches Museum von Meisterwerken der Naturwissenschaft und Technik (German Museum of Masterworks of Natural Science and Technology). A central feature was the Ehrensaal (Hall of Fame) in which portraits of famous German scientists and engineers were displayed. The term 'masterwork' in the Museum's title provided the criterion by which historical objects were selected. Miller's determination that his Museum should demonstrate the interactions of science, technology and industry during their historical development; his effective use of working models for the demonstration of scientific and mechanical principles; his belief that the Museum should entertain; and his obvious success at achieving all of these, in turn influenced the policies of the museums he had visited and provided inspiration for the establishment of others.

In the United States the Smithsonian Institution had collected material relating to the physical sciences and technology under the auspices of the National Museum, without any real mandate to do so, but the Museum's participation in the 1876 Centennial Exhibition in Philadelphia and in particular its designation as the official repository for the objects exhibited there, was to lead directly to the opening in 1881 of the Arts and Industries Building. But despite a series of initiatives culminating in a vigorous campaign in the 1920s in which it was stated that the nation needed 'a South Kensington, and Deutsches Museum rolled into one as befits its size and wealth', no free-standing museum of science, engineering or industry was established in Washington.

The key figures in these efforts had been Carl W Mitman (1890–1958), Head of the Department of Mechanical and Mineral Technology in the Smithsonian, and a New York consulting engineer Holbrook Fitz-John Porter (1858–1933). They invoked the patriotism of American engineers, appealed to their professional pride, obtained official backing from the societies of mechanical, mining, civil and electrical engineers – although no promises of funds – and commissioned an architect to prepare a scheme for a National Museum of Engineering and Industry, to be the largest museum in the world. A fund-raising committee included such luminaries as Thomas Edison, Orville Wright and Julius Rosenwald, the philanthropic head of Sears Roebuck & Company.

Throughout the 1920s Mitman and Porter campaigned. It was a period of scientific and industrial renaissance in the years after the war to end all wars. Industrial prowess and national power were seen as synonymous. In Germany and Britain great new purpose-built national museums of science and industry were nearing completion, to open in Munich in 1925 and South Kensington in 1928.

The theme of national achievement within an international arena had for long been popular. In 1849, prior to the Buckingham Palace Conference of 30 June, Henry (later Sir Henry) Cole (1808–82), a member of the committee set up by Prince Albert to plan the Hyde Park exhibition, had hinted to the Prince that the works of all nations might be included. At a stroke the concept of a national exhibition had taken on a new and visionary dimension to be realized in the Great International Exhibition of 1851. Thenceforth a succession of exhibitions, expositions and world fairs were to provide the backdrop for the presentation of indigenous achievements in the setting of a great international event.

And it was to a great extent by the engineering and industrial exhibits that the power and influence of the great nations of the world were measured. Mitman and Porter tried without success to exploit not only the emergent anxieties in America's engineering community about their position within the nation's culture but, by stressing the benefits that experiential knowledge gained in the learning laboratory of a museum might bring, they sought to tap new but ill-formed concerns about science and technology education.

Their campaign touched other raw nerves too, such as those of entrepreneurs whose fortunes had been made in engineering but who bought immortality by endowing art museums. H W Dickinson, then Keeper of Mechanical Engineering at the Science Museum in London, in offering persuasive and eloquent support, said, 'one would conclude that it was almost indecent to engage in industry, but having done so, and then been successful, only a peace offering to the Muses would suffice as reparation'. His examples were steel-maker Samson Fox, patron of the Royal College of Music, and Sir Richard Tangye whose money made out of engineering had supported Birmingham City Art Gallery. Porter quoted Detroit manufacturer Charles L Freer who endowed the Smithsonian's Freer Gallery of Art.

Mobilizing support within the cultural establishment of the Smithsonian proved impossible; and then came the Depression. It was to be over thirty years before a new museum, approximating only in

The Crystal Palace,
completed in 1851 to house the Great Exhibition of the
Works of Industry of all Nations

the most generalized sense to what Mitman and Porter had in mind, came into being. The National Museum of History and Technology opened in 1964. Here science and technology were set in a political, economic and cultural context. This was almost the museum South Kensington might have had if art and science had not gone their separate ways at the turn of the century. The change of name to the National Museum of American History in 1980 symbolized the further integration of science, technology and industry and the final extinction of dreams for a free-standing celebration of engineering heroism and industrial power. But the United States was to gain that with the opening in 1976 of the National Air and Space Museum, the world's largest and most visited museum. Its direct equivalent in Britain is the National Railway Museum, York, which had opened the previous year. Both museums celebrate the great gifts to humanity of the steam railway and powered flight.

Although Mitman's dream was never realized he does have his memorial, like Bennet Woodcroft's collection of portraits of engineers, in over 300 biographical entries on American technologists, almost all engineers, in the *Dictionary of American Biography*. But the failure to realize Mitman and Porter's dream in Washington in the 1920s was not the end of the story. Julius Rosenwald (1862–1932) returned to Chicago where the Palace of Fine Arts, the only significant structure surviving from the World's Columbian Exhibition of 1893, still stood empty and decaying. Rosenwald funded its complete rehabilitation and it reopened as the Museum of Science and Industry in time to receive many of the exhibits from the next Chicago World's Fair, *A Century of Progress*, in 1933–4. Meanwhile, Orville Wright had presented the famous Kitty Hawk Flyer, in which he and his brother Wilbur made the first powered flight in 1903, to the Science Museum in London in 1928. It was returned to the United States in 1948 and now holds pride of place in the National Air and Space Museum. Paradoxically it was also in 1928 that Edison's original phonograph, acquired in 1880 by the Patent Museum, was sent back to the United States as the result of a direct request through the British Ambassador in Washington.

The Science Museum's acquisition of seminal objects from the United States and elsewhere – the Haber-Bosch ammonia synthesis pressure vessel from Germany is a recent example – is reflected in an even greater enthusiasm by museums overseas to record in their collections the heroic period of British industrial and engineering innovation. The Henry Ford Museum and Greenfield Village in Dearborn, Michigan, for example, opened in 1929 by Ford (1863–1947) as a tribute to his hero Thomas Alva Edison (1847–1931), was designed to portray the United States during its transition from a European pre-industrial tradition to a technology-based society, and provide through a collection of engines and mechanical devices an illustration of the march of engineering progress. An essential prerequisite was for a comprehensive collection of stationary steam engines and this was assiduously assembled by Ford's team of engineers in Britain during the mid-1920s, long before the growth of industrial archaeology was to accord them any greater cultural significance. (In recent years the Science Museum has assisted in the restitution of one of these engines collected by Ford from Coalbrookdale, Shropshire, and now returned there). The Deutsches Museum too recognized the need to establish the credentials of early industrialization but made do with replicas, faithfully reproducing, among others, the Boulton and Watt rotative beam engine and *Puffing Billy*, both in the Science Museum's collections.

Elsewhere, museums of industry, engineering and transport reflect the enormous output of equipment and goods manufactured in Britain to serve the needs not only of her growing empire but a wider world of which for a period of perhaps three or four generations she was, quite literally, the workshop. From Gävle in Sweden to New Delhi, from Brisbane to Buenos Aires, the products of British railway engineers dominate museum collections. The genius of Port Glasgow, Elswick, Belfast and Birkenhead is preserved in Honolulu, Yokosuka, Melbourne and Talcahuano. In Nagoya the museum of the Meiji period immortalizes Japan's debt to Birmingham, Manchester, Glasgow and Newcastle.

These products of British industrial enterprise dispersed throughout the world are preserved more extensively than the material evidence of any other cultural tradition. This great global collection is not simply a homage to engineers and their skills, nor to imperial power or merchant entrepreneurs. It marks above all the first stages of the universal technology on which the contemporary world is based.

Today science and technology know no boundaries. Discovery and invention can rarely be ascribed to one country or even continent let alone an individual or corporation. Research and development and the manufacturing processes that are their outcome take place on an international stage, commanding skills and investment simultaneously from many places. The collections of the Science Museum track the origins and progress of this new world order.

MILESTONES OF
SCIENCE AND TECHNOLOGY

The following pages show one hundred
key developments in science, technology and medicine.
Their purpose is to demonstrate to as wide an audience as possible
the creative achievements that have been made from antiquity
to the present day and which shaped
the modern world

BYZANTINE SUNDIAL-CALENDAR

We live surrounded by mechanisms, relying on different ones almost for each moment of the day. If we pause to consider the origins of all these devices we may be surprised to realize that mechanical engineering as we know it is a very recent profession, and its origins are surprisingly obscure.

A widespread feature of much mechanism is the use of toothed gearwheels, and this early Byzantine sundial-calendar acquired and identified by the Museum in 1983 contains what are thought to be the second oldest gears in the world. Incomplete as the device now is, and simple as it probably was when complete, it is of immense historical importance.

The world's oldest geared mechanism, known as the Antikythera Mechanism after the island near which it was recovered from the sea, is in the National Archaeological Museum in Athens. Dating from the first century BC, it contains an astonishingly complex assembly of over thirty gears, the purpose of which is not wholly understood. The inscriptions and the arrangement indicate connections with astronomy and with the calendar, but current work suggests that the reconstruction needs to be revised.

It had long been supposed that the Antikythera Mechanism was the sole representative of a Hellenistic tradition of geared devices which, sometime in the ensuing thousand years, was transmitted to the Islamic world. The next known evidence for geared mechanism after the Antikythera Mechanism was an account by the Persian polymath Al-Bīrūnī in about AD 1000. The 'Box of the Moon' was a gadget in which gears connected four displays showing days, age of the Moon (days since last New Moon), and positions of Sun and Moon in the zodiac.

The discovery of this Byzantine instrument gives us a reference point in the middle of that thousand-year interval. As with the Antikythera Mechanism, its Greek inscriptions place it in the world of Hellenistic culture; on the other hand the actual arrangement of the surviving wheels corresponds exactly to that described by Al-Bīrūnī, demonstrating the strong probability that Al-Bīrūnī's device was a direct borrowing from the Greek tradition. Thus this object at once adds significantly to the surviving evidence of ancient technology in general, and provides the first direct evidence for the transmission of mechanical technology from Hellenistic to Islamic culture.

From the fragments it is clear that this early Byzantine sundial-calendar has two apparently distinct functions. A large disc is recognizable as the basis of a portable sundial. An arm with a swivelling ring is

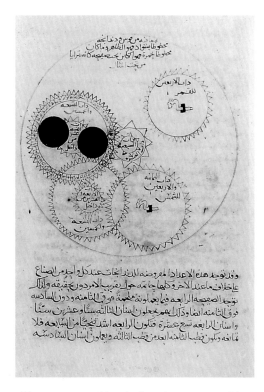

The arrangement of the wheels in Al-Bīrūnī's 'Box of the Moon', from one of the earliest known manuscripts of his work, dated AD 1217. The central wheels and the pair with the black discs correspond to surviving parts of the Byzantine Sundial-Calendar (University of Leiden Library)

connected with its use as a sundial, because in use the sundial had to be suspended to hang upright, but the shape of the arm shows that unlike the other known examples the body was a circular box, not just a disc. That box contained a geared calendrical mechanism very like that described by Al-Bīrūnī, which makes this instrument unique. Heads engraved in a circle on the large disc are characterizations of the days of the week, and the piece that fitted through the hole in the centre of the circle has also survived; this piece was turned by hand, perhaps by putting a key on a projecting square. Another surviving part is geared to it, providing a visual display of the phase of the Moon. Similar displays are found on some modern wrist-watches today, and this is the oldest known example!

That is almost as far as the fragments can take us, except that each of the rotating pieces carries another gear wheel that has to be explained. The Museum's reconstruction was based on accounting for and making sense of the surviving features by adding as little as possible, in the simplest possible style. This entailed following almost exactly the arrangement of Al-Bīrūnī's Box of the Moon, with displays of the positions of the Sun and the Moon in the zodiac.

Intriguing questions remain as to whether the original instrument might have been more complicated. For example, by a fairly simple addition to the gear system, the calendar could be made into a predictor of lunar and solar eclipses. By the further addition of scales or tables on the (missing) back of the case, to assist with the tricky conversions entailed, the sundial and calendrical parts could be used together to tell time at night, by the Moon. But fortunately the importance of this rare survival does not depend on the details of reconstruction; the existence of these rare fragments and our ability to ascribe such an early date to them are enough to place them amongst the Museum's treasures.   MTW

ISLAMIC GLASS ALEMBIC

An alembic is the essential piece of distillation apparatus used by early chemists. The liquid to be distilled is heated in a lower vessel and the resulting vapour condenses on the inside of the dome shape of the upper vessel, the alembic. The condensed vapour runs down into the internal gutter and out through the spout into a collecting vessel.

Distillation was one of the arts developed by the Alexandrian alchemists (AD 100–900) and contemporary illustrations, although rather crude to our eyes, show the basic shape with the internal gutter and long spout. At this early date, the process was carried out on a small scale for the purification of liquids and the preparation of medicinal essences.

The traditional skills were adopted and developed by Arab chemists, but the apparatus remained essentially the same, as indeed it does today. The Persians were particularly noted for their use of distillation to produce essential oils and perfume. This glass alembic, probably dating from tenth- to twelfth-century Islam, is thought to be one of the oldest to survive from the West or Near East.

Distillation is one of the oldest chemical processes known. By the Middle Ages it was in use for the preparation of strong acids and also – perhaps a more familiar use – for making alcoholic drinks. In the late nineteenth century, the process was developed on an industrial scale for the distillation of petroleum and is still an important technique in the chemical industry.

In view of the importance of the process and its continued use from early times, it is not surprising that the Science Museum includes this alembic as

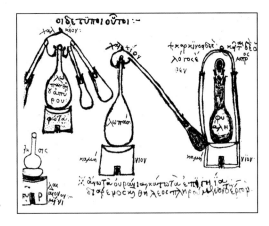

A third-century AD illustration of alchemical distillation apparatus showing (left) two alembics

one of its prime objects, but it was largely a matter of luck that the group of vessels including the alembic was acquired.

An entry in the catalogue of a sale of early glass caught the attention of a colleague from another museum, an authority on early glass. Consultation with the Science Museum's Keeper of Chemistry in a breakfast-time telephone call confirmed his suspicion that the item described as 'A fine and rare Roman glass spouted funerary vessel' was indeed an alembic photographed upside down.

The vessel had apparently been found with the next lot in the sale, described as 'A fine and rare Roman glass funerary vessel and cover'. This was thought to be a sublimation apparatus (for distilling solid substances) and was another beautifully made

piece of early chemical glassware. It consists of a large vessel with a lid which fits well and the decoration matches, so they seem likely to have been made to fit together.

Inspection of the items at the saleroom confirmed their opinion, although unfortunately no one knew where any of the pieces originated from. The distinctive shape of the alembic leaves no doubt as to its identity and it was described by the Keeper as 'one of the rarest objects I have ever seen'. Furthermore, the surface weathering is typical of Islamic glassware of the tenth to twelfth centuries. The spectacular iridescence on the surface of the glass is caused by interference patterns between layers of weathered glass and the air trapped between them.

Glass making flourished in the Islamic period and the glass has a good record of survival due to its chemical composition. The basic raw materials for glass making are silica, lime and an alkali. In the coastal regions of the Mediterranean, marine plants which yield the sodium based compound, soda, were used as the source of alkali. This constituent is largely unaffected by water and gives a glass which remains stable with time.

By contrast, the glass made in the forest glass houses which flourished from medieval times in inland regions, includes alkali made from beech or other hardwoods. This gives potash containing potassium which is readily attacked by water. When buried these glass artefacts disintegrate quickly; fragments of distillation apparatus from a fifteenth-century monastic site at Selborne in Hampshire are barely recognizable as glass.   SJC

THE GIUSTINIANI MEDICINE CHEST

Imagine a three-month voyage on a galley in the eastern Mediterranean. Imagine falling ill with fever, dysentery or delirium and being cured by aromatic water of viper's bugloss, powder of camomile, preparation of stag's horn or metallic sulphur. The world from which this magnificent chest survives is so remote that it is difficult to empathize with the owners of what was, in effect, a huge, seagoing first aid kit.

Solidly constructed from partly gilded wood, its leather covering secured by gilded nails, the chest measures some 36 cm × 44 cm × 68 cm and takes two men to lift it. The hinged lid is secured by a single warded lock and opens to reveal, on the underside, a much retouched painting on canvas of a female figure in a classical landscape. Beneath the lid, an upper compartment and three velvet lined lower drawers, which open to right and left, contain 126 drug bottles and pots.

These drugs form a microcosm of Renaissance pharmacy; a time capsule of medicines from the sixth decade of the sixteenth century, for the chest can be dated with a high level of certainty. The chest was apparently made as a sea medicine chest for Vincenzo Giustiniani, last Genoese ruler of the Aegean island of Chios. Vincenzo ruled Chios from 1562 until it fell to the Turks in 1566. Subsequently he became Charles II of France's ambassador to Istanbul. The arms of the Giustiniani of Chios are crudely painted on a wooden box housing scales and brass weights in the middle drawer of the chest, thus indicating a date of 1566 or earlier.

The sixteenth century saw two major changes in Western pharmacy. From antiquity, simple plant extracts had been used to make up medicines based on instructions given in the famous herbal of Dioscorides, written in Greek in the first century AD. New translations of the herbal had been published at Paris and Venice in 1516 and reached a far wider circulation than earlier manuscript versions. At the height of Dioscorides' influence, however, came new drugs from a New World: America. And later in the sixteenth century the remedies of the German physician, Paracelsus (1493–1541) began to be adopted. Paracelsus abhorred polypharmaceutical preparations (drugs mixed from a number of medicines), and introduced chemical drugs, such as sulphur and mercury, for the first time. Ninety-five of the chest's containers have paper labels with legible or partly legible names, apparently written in a sixteenth-century hand. Of these, thirty-nine have been identified by modern scholars and almost two-thirds of these are mentioned in Dioscorides' herbal. At least two others, guaiacum and mechiocam, are American drugs and another, a preparation of antimony known as *crocus metallorum*, was first described by Paracelsus. Guaiacum wood had been known in Europe since 1517 and was used as a remedy for the syphilis which raged there during the sixteenth century. The chest also contains extremely costly remedies, such as *terra sigillata* (sacred sealed earth from, in particular, the island of Lemnos, or sometimes Malta or Jerusalem) and bezoar stone, and at least one preparation (for preventing miscarriage) specifically for women.

The Giustiniani chest is certainly unique. It has been described as the earliest extant sea medicine chest but it differs considerably from other sea chests. From the sixteenth to the nineteenth century these were workmanlike kits, with instruments and bulk supplies of everyday ointments and medicines used by surgeons to treat injury and disease amongst the crew. The nearest contemporary sea medicine chest known, that of the Tudor warship, *Mary Rose*, which sank in Portsmouth Harbour in 1545, would appear from its remains to have been far less sophisticated in content than the Giustiniani chest. The Giustiniani chest is perhaps better regarded as the personal medical chest of a wealthy sixteenth-century family whose members would not be without it, even at sea.

GML

THE STANDARDS OF THE REALM

Shared standards of weight and measure lie at the heart of each organized and advanced society, promoting fair and peaceful trade, and planned construction of buildings and shipping. The search for such standards goes back to ancient times. Early people related measure to parts of the body, thus the cubit (measured from the elbow to the finger tip) and the foot. The Romans adopted various weights and measures from the lands they conquered, but certain standards spread throughout their great empire were used in England during the period of occupation, and left their mark on subsequent practices.

In England other influences, such as those of the Anglo-Saxons, were brought to bear; as trade grew more measures were adopted and local variations made the system complex and open to abuse. Successive medieval monarchs issued laws to promote fair and standard measures, but enforcement was a substantial problem. Moreover, where standard measures were supplied for reference, great care had to be taken to ensure they were made accurately and then were well maintained. Error could all too easily creep in if standard measures were broken or succumbed to wear and tear.

With the Tudors came a new wave of energy. Under their rule came rapid economic growth in agriculture, commerce and industry. Tudor monarchs understood the need for good commercial administration and began a vigorous campaign to establish standard weights and measures. Under Henry VII, new Exchequer standards of weight, length and capacity were made, and copies of them were sent to forty-three shire towns in England. Amongst the earliest English measures still in existence, are two of Henry VII's capacity measures, the standard gallon and the standard bushel, issued in 1495.

More significant advances were made under Elizabeth I although it proved difficult initially to establish accurate standards. A first series of weights was issued in 1558, but proved to be too heavy. In 1574, responding to complaints that the weights in use varied too greatly, the Queen set up a jury of merchants and goldsmiths to produce new standards. Those they produced were also faulty. A second jury produced a new set of primary reference standard weights for the Exchequer in 1582. During the next six years, fifty-seven sets of careful copies were made, to be sent to the major towns and cities of the realm. There they would serve as reference to be used by all local traders. Sets were also sent to the Tower of London and to the Goldsmith's Company in London. Weights that did not match the standards were to be destroyed and a proclamation was issued calling for nationwide use of the new standards.

The Science Museum has a splendid collection of these standard weights from the reign of Elizabeth I. They are made of bronze and are engraved with the royal insignia and the dates 1582 or 1588. The troy weights (used to weigh coin or bullion) are a series of nested cups weighing from one-eighth of an ounce to 256 ounces. The avoirdupois standards (used for all other general weighing) take the form either of bell-shaped weights, with looped handles, or of a series of circular discs with slightly raised rims to allow the weights to be stored one atop the other. These range in weight from two dram (one-sixteenth of an ounce, avoirdupois) to eight pounds.

Whereas standards produced by previous mon-

archs generally became inaccurate and had to be withdrawn, those of Elizabeth I were so good they remained in use until 1824.

Standards of length were also issued during the Tudor era: Henry VII issued a standard yard in 1497, an octagonal measure made of bronze. The Exchequer standard yard issued by Elizabeth I in 1588 was based on this. Also issued in 1588 was a standard ell, used for cloth, with a length of forty-five inches. Both these bronze measures remained in use until 1824, from which date the ell ceased to be a legal measure.

Having issued standard weights and measures of length, in 1601 Elizabeth I issued standard capacity measures to sixty cities in England and Wales. The bushel and gallon were to the same standard as those issued by Henry VII. The Science Museum holds a group of fine capacity measures of Elizabeth I: the standard bushel, two standard gallon measures, a quart and a pint. All are made of cast bronze with handles to use when lifting. Again, these were so well made that they were in use until 1824. The Tudor standards improved greatly on what had gone before and gave long service through a period in which Britain's might as a trading nation grew. There were still difficulties in enforcing the weights and measures system consistently; too great a variety of weights and measures remained in use (some in particular trades, some in particular localities) and legislation governing this important area was too varied to be clear and comprehensive. Yet the Tudors had made unprecedented efforts to establish and maintain a nationwide system of benefit to all, at a time when the majority of countries in Europe tended to use purely local standards of measure.   PR

SLIDE RULE BY ROBERT BISSAKER

Until the early 1970s the slide rule was a familiar and established tool for calculation. Although its use was taught in many schools, it was most heavily identified with the professional world of engineering. Indeed, earlier in this century the 'slipstick' slide rule was virtually a symbol of the engineer, who would have it always on hand to calculate results quickly where the highest level of precision was not required.

However, the widespread educational and engineering use of slide rules in the twentieth century disguises a much longer and richer history, for slide rules were first devised in the seventeenth century. This example is the earliest dated straight slide rule (1654), the most familiar of the several forms in which these devices have been made. The crucial ingredient is a set of logarithmic scales, so any account of slide rules must begin earlier in the seventeenth century with the invention of logarithms by the Scottish mathematician John Napier (1550–1617).

This new method of calculating was first described in print in Napier's *Mirifici Logarithmorum Canonis Descriptio* (*A Description of the Marvellous Canon of Logarithms*, 1614). The logarithms were taken up most quickly and actively by the 'mathematical practitioners' of contemporary London. An English translation of Napier's first text rapidly appeared, while logarithms were also mathematically reformulated in a new and more practical version. The integration of logarithms into the existing repertoire of calculating devices and practical mathematics was sealed by the English mathematician Edmund Gunter (1581–1626). Gunter published his account of a calculating instrument using logarithms in 1623, although the arrangement of his logarithmic scales was not a slide rule as such. Rather than providing pairs of scales which would be slid against each other, Gunter described a set of single scales. To use these a pair of dividers or compasses was required, with which the quantities were stepped out along the length of the scale.

The announcement of logarithms; frontispiece of
John Napier's *Mirifici Logarithmorum Canonis Descriptio*, 1614

Gunter's logarithmic scales were developed in a multitude of forms – straight, circular and spiral – by a whole host of mathematical practitioners and instrument makers. The first instrument which can be called a slide rule was made public not as a straight rule but as a circular device, and considerable controversy surrounded its invention. The English mathematician William Oughtred (1575–1660), who has the best claim to have invented the circular slide rule, also appears to have been responsible for the straight rule: he achieved this by the simple expedient of placing two Gunter scales side by side.

Bissaker's straight slide rule bears essentially the same scales that Gunter had recommended more than thirty years before, though like Oughtred's its use does not require the dividers which are an indispensable accessory to Gunter scales. Compared to other surviving instruments made of wood (as opposed to the more expensive brass, or even silver), Bissaker's rule is unusual in having a relatively informative inscription by the maker:

* MADE BY ROBERT BISSAKER * 1654 * FOR T W

However, despite this information, Robert Bissaker for long remained no more than a name. We now know that Bissaker was an instrument maker who belonged to the Stationer's Company of London. He was of Shropshire yeoman stock and came to London in 1634 as an apprentice to one Nathaniel Gosse, who must have taught him the skills of the instrument maker's trade. In 1642 Bissaker was to be found making wooden instruments 'at Rat-cliff, over against the Red-Lion Tavern', and there he seems to have remained into the 1660s. He presumably catered largely for a market of seamen and navigators since Radcliffe was a small but important town on the Thames, rapidly being engulfed by the eastward expansion of London.

From their very early years onwards, particular applications were found for slide rules. New designs adapted for navigation, surveying, excise calculations and carpentry were supplemented in the nineteenth century by a plethora of specialized uses, from cattle weighing to chemistry. Despite the availability of numerical tables and mechanical calculators in the nineteenth and twentieth centuries, slide rules remained invaluable, and by the middle years of the present century they could be found in almost any activity requiring calculation. It was not until the advent of small electronic calculators that slide rules were comprehensively challenged and displaced: not for nothing were the first hand-held scientific calculators known as 'electronic slide rules'.

While never trumpeted as a grand or radical instrument destined to change the world, the portability, speed and convenience of the slide rule conferred a cheap but powerful means of calculation on its owner. As made by Robert Bissaker the slide rule was a humble object, yet although lacking the glamour of more obviously heroic objects of discovery and invention, slide rules have been just as essential to generations of engineers, scientists, and numerate professionals.   SAJ

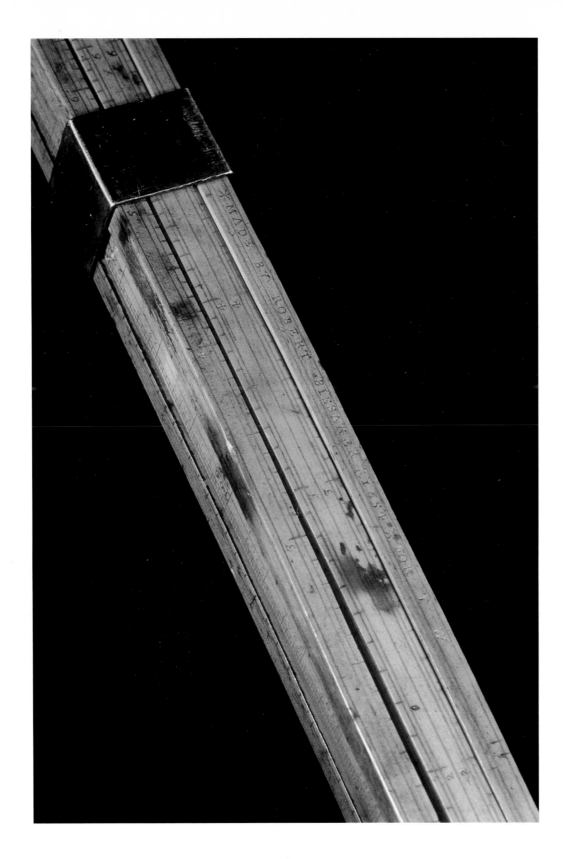

NAPIER'S BONES

John Napier (1550–1617), Baron of Merchiston, is renowned for his invention of logarithms – a contribution to science viewed by some as ranking in importance second only to Isaac Newton's incomparably influential *Principia*. One of the practical advantages of logarithms is that their use allows lengthy multiplication and division of numbers to be performed using simple addition and subtraction only. The process of looking up values in printed mathematical tables and doing simple arithmetic not only simplified arithmetical tasks but was less liable to human error than conventional longhand methods.

Napier's logarithms also provided the basis for the first slide rule (*See* SLIDE RULE BY ROBERT BISSAKER). In about 1620 Edmund Gunter plotted logarithms on a straight scale and performed multiplication and division by adding or subtracting lengths using a divider. The leap from Gunter's 'line of numbers' to the slide rule was made by William Oughtred as early as 1621.

Napier was responsible for another less significant but highly intriguing physical aid to calculation known variously as 'Napier's Bones', 'Napier's Rods', or 'Speaking Rods' though the first of these has stuck probably because of its macabre overtones. However, the 'bones' do not refer to relics of Napier's skeleton but to the fact that the more expensive versions of the device were made from bone, horn or ivory. Napier published his description of the use of the bones in 1617, the year of his death.

The bones, devised to aid multiplication and division, consist of numbered rods or oblong blocks. Each face of the rod is marked at the top with one of the ten counting digits and below are listed each of its multiples. The rods are laid alongside each other so that the multi-digit number to be multiplied appears in the topmost row. Any multiple of this number can then be read off right to left along the row of the required multiple. The trick of the bones is in the layout of the numbers. The units and tens are separated by a diagonal, and in a given row the tens digit from a column on the right shares a

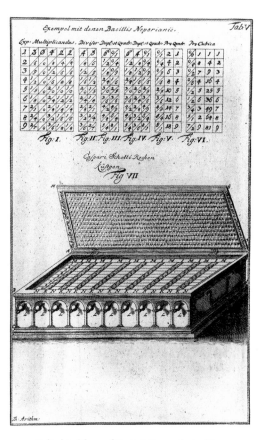

parallelogram with the units digit in the column immediately left. By reading from right to left and mentally adding pairs of numbers in each parallelogram on the same row, answers can be read off directly and written down. For example, to multiply 272,968 by 4, run down the left-hand index to '4' and read that row from right to left adding number pairs, ie, 1,091,872.

A set of bones is in effect a multiplication table sliced up into moveable columns and the arrangement reduces multiplication to a series of lookup operations and simple additions. Division was assisted by using the bones to perform trial multiplications and separate sets of bones were devised to calculate square and cube roots.

Variations and extensions of Napier's Bones took several forms. Sets were devised for special purposes including geometry, planetary movements and astronomy. Different physical arrangements were devised: there were versions with rotatable cylinders instead of oblong strips, and at least one version with circular discs. A final development took the form of a series of rulers devised by Henri Grenaille and demonstrated in 1891. Grenaille's layout removed the need for any mental effort in performing the 'carry' when reading off products.

Napier's Bones were a great vogue and their popularity can be seen as evidence of the low standards of numeracy in the early seventeenth century even among the well educated. Versions of varying degrees of plushness were produced – numbers carved in ivory in a leather case, for example. The bones eventually lost favour as familiarity with techniques of arithmetic began to spread.   DDS

Cylindrical form of Napier's Bones, devised by
Gaspard Schott, 1668. The cylinders are rotated by
turning the knobs on the front of the box.
Results are read off from the cylinders through slits,
each of which reveals one column of figures

HAUKSBEE'S AIR PUMP

An air pump – today we would call it a vacuum pump – is a device for extracting air from a vessel to produce a partial vacuum. In the eighteenth century its importance lay in the fact that a vacuum was unexplored territory, a medium in which new experiments could be tried and unusual phenomena observed. Early pumps had been developed by Robert Boyle and others, but Francis Hauksbee (c 1666–1713) so improved their design that his pumps were said to be the best in the world.

Francis Hauksbee first demonstrated his 'New Invented Air Pump' at a meeting of the Royal Society in London in December 1703, presided over by Isaac Newton who had been elected President of the Society a few weeks earlier. The Society was a forum where men could meet to share an interest in 'natural philosophy'. An important aspect of the Royal Society was that new ideas were demonstrated rather than just talked about. 'There's no other way of *Improving NATURAL PHILOSOPHY*', wrote Hauksbee in 1709, 'but by *Demonstrations* and *Conclusions*, founded upon *Experiments* judiciously and accurately made.'

Hauksbee's first demonstration of the air pump so captivated the fellows of the Royal Society that they soon appointed him official demonstrator at all their meetings. As a professional instrument maker, he was highly proficient in the design and construction of experimental apparatus. He was also one of the first to bring natural philosophy to a wider audience by offering public lecture courses, open to anyone willing to pay the fee. Over the years his skilfully staged experiments and demonstrations were to provide much food for thought for Newton and his contemporaries as well as the general public.

This pump must have been made by Hauksbee some time before 1709, as he used it to illustrate his book *Physico-Mechanical Experiments on Various Subjects*, published the same year. The principle is similar to that of a bicycle pump, adapted to suck rather than blow (it takes the form of two such pumps side by side). The experimental apparatus is placed inside the vacuum vessel (o), which is connected by a

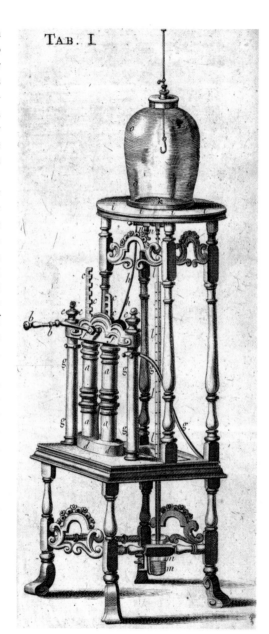

Engraving of the pump in Hauksbee's
Physico-Mechanical Experiments, 1709

narrow tube (h) through valves to the base of two cylindrical barrels. When the handle (b) is turned clockwise, a rack and pinion mechanism raises the piston (c) in the left-hand barrel, drawing air from the vessel down into the barrel. At the same time the right-hand piston moves down, expelling the air inside it which had been drawn out of the vacuum vessel on the previous stroke. After a few turns clockwise the direction of the handle must be reversed and the process repeated, with the two barrels exchanging roles. Air from the vacuum vessel is now drawn into the right-hand barrel, while air already in the left-hand one is expelled. The whole cycle must be repeated until the degree of vacuum in the vessel, indicated by a mercury barometer (m and l), reaches the required level, or until a stage is reached at which the inevitable small leakage of air back into the system balances the rate at which air can be pumped out.

A wide variety of apparatus was available for use with the pump. Some of the simpler demonstrations of Hauksbee's time, such as the one to show that the sound of a bell reduces as the degree of vacuum increases, remain in the schoolroom repertoire to this day. Others, involving living creatures, are thankfully no longer employed. Hauksbee himself made a particular study of the way electrified objects behave in a vacuum, finding that a spinning glass globe, with the air pumped out from inside it, gives out a purple glow when rubbed with the hand. Similar experiments, done more than a century later when new sources of electricity and improved vacuum pumps were available, would eventually lead to an avalanche of discoveries which included X-rays, the electron, and the television picture tube.

The Hauksbee air pump carries special status. It links us directly with a period in history when the idea that nature's secrets could best be unravelled by purposeful experiment was a new one. How extraordinarily fruitful that idea was to prove in the centuries ahead, neither Hauksbee nor any of his contemporaries could have foreseen.   AWW

THE ORIGINAL ORRERY

Devices which enlarge the microscopic or reduce the cosmic to a manageable human scale enjoy a broad appeal. It is not only scientists who wish to make models of the solar system – the Sun and its family of nine known planets and numerous natural satellites.

Since the Middle Ages planetary machines have been constructed, mostly by clockmakers, to show the relative positions and apparent motions of Sun, Moon and (sometimes) planets. Some of these planetary machines revealed astonishing complexity and sophistication in their construction, among them Giovanni de Dondi's astrarium of 1364, Philipp Imsser's astronomical clock of 1555, Eberhard Baldewein's planetary clock of 1561 and Christiaan Huygens' planetarium of 1682.

The orrery entered the scene as an altogether simpler planetary machine, and hence it was made in greater numbers and was more widely distributed. It is a partial model of the solar system as seen from the outside – a 'god's-eye' view – with the Sun at the centre, following Copernican principles. In its earliest and simplest form, the orrery showed no planets other than the Earth revolving about the Sun plus the Moon revolving about the Earth. It is in this form that the English antiquarian William Stukeley (1687–1765) described devices made (possibly) by Richard Cumberland and Stephen Hales. But the first serious examples were made by the celebrated London clockmaker George Graham (c 1673–1751).

Graham probably made his planetary machines, two of which are known to survive, between 1704 and 1709 when he worked as senior journeyman under that other great clockmaker Thomas Tompion at the Dial and Three Crowns in Fleet Street. Another Fleet Street maker, John Rowley (c 1668–1728) at the Globe, saw Graham's planetary machines and copied one of them on a commission from Charles Boyle, fourth Earl of Orrery (1676–1731), among whose ancestors was the famous natural philosopher Robert Boyle. Although by rights the new planetary machine could well have been called a 'graham' after its inventor or a 'rowley' after one of its earliest makers, it was in fact named an orrery after its purchaser and patron.

This planetary machine was the first to be called an orrery and Rowley is thought to have made his example for the Earl of Orrery around 1712. Whereas earlier astronomical mechanical devices were generally unique objects destined for wealthy patrons, orreries were to become relatively cheap and popular in the course of the eighteenth century. William Jones' small portable orrery, for example, 'recom-

Another orrery by John Rowley
from Nicolas Bion's *Traité de la Construction et des Principaux Usages des Instruments de Mathématique*, translated by Edmund Stone, 1737

mends itself to the Public for Simplicity and Cheapness, particularly to Masters and Governesses of Boarding Schools, Private Tutors, &c'.

The original orrery is shaped like a drum with a diameter a little smaller than that of a modern bass drum. Under the central glass dome, the brass ball of the Sun rotates as it ought in twenty-seven and a half Earth days. The top surface of the drum, decorated with random white stars on a blue ground, bears the Earth-Moon system under a second glass dome. One turn of the hand crank on the rim of the drum rotates the ivory sphere on the Earth exactly once – a full twenty-four-hour day. The Moon revolves about the Earth in a lunar month of twenty-nine and a half Earth days, whilst rotating on its own axis so as to keep the same face directed towards the Earth.

Mounted on wooden pillars around the periphery is a flat brass ring representing the plane of the ecliptic – the Earth's path around the Sun or, equivalently, the Sun's apparent path through the constellations of the zodiac. The scale on the ecliptic ring, against which a pointer attached to the rotating star-plate indicates the date, includes the symbols of the zodiacal constellations. Inside the drum are the gear-trains which control the solar, terrestrial and lunar motions.

The accuracy of Rowley's gear-cutting is amply reflected in the orbital motion and axial rotation of the three celestial bodies shown. What the orrery cannot be expected to model correctly is relative size. For the Earth to be in the right relation to Rowley's three-inch Sun, it would have to be the size of a grain of sand a bus-length away.   JRD

SISSON'S RULE

At the sale of the instruments of the late ingenious optician Mr. JAMES SHORT, I purchased a finely divided brass scale, of the length of 42 inches, with a VERNIER's division of 1000 at one end, and one of 50 at the other, whereby the 100th part of an inch is very perceptible. It was originally the property of the late Mr. GRAHAM, the celebrated watch-maker; has the name of JONATHAN SISSON engraved upon it; but is known to have been divided by the late Mr. BIRD, who then worked with SISSON.

So wrote Major General Roy in 1785. At the time, the Royal Society had in its possession another brass standard scale forty-two inches long, marked with the length of the standard yard of the Realm from the Tower, the standard yard from the Exchequer, and the French length of a half-toise, also made by Jonathan Sisson (fl. 1736–88). Major General Roy (1726–90) was anxious to confirm that his scale matched the Royal Society one because he intended to use his scale to prepare some of the measuring equipment for setting out a base line on Hounslow Heath. This work was a necessary preliminary to fixing the relative positions of Greenwich and Paris observatories with greater accuracy than ever before.

The principle was simple: if the length of the baseline was known, the distance between each end of the base line and a distant landmark could be worked out using a theodolite (See RAMSDEN'S THREE-FOOT THEODOLITE) to measure the angles of an imaginary triangle formed by the base line and the landmark. A simple trigonometrical calculation gave the distances. From Hounslow Heath a web of triangles was surveyed across Southern England towards Dover and a second base line at Romney Marsh, and then simultaneous sightings across the Channel linked the English measurements to those of the French. The work was carried out at the behest of the Royal Society of which Roy was a

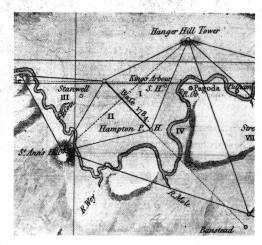

Map showing the first part of General Roy's triangulation of 1785. The pattern of triangles shows how the combination of an accurately laid-out base line and exact angles measured by theodolite sightings on landmarks gave a network of triangles from which distances and bearings could be calculated

Fellow, and the French Académie des Sciences. Although Roy died in 1790, his work became a model for the extension of the project to map the rest of Britain.

On 16 April 1784, a select group of scientists from the Royal Society convened at Hounslow Heath for a preliminary survey. The heath was ideal; it had an extremely level surface, was large, near both London and Greenwich, and with few obstructions likely to make measuring difficult. The team readily settled upon a line running for about five miles from the north-west extremity of the heath towards the south-east, and agreed that soldiers should be employed to clear furze bushes and ant-hills along the track of the base, and to assist with the measuring operations.

It took until early July to clear the route for the line, during which time Jesse Ramsden (1735–1800),

the instrument maker, prepared the instruments. The Sisson Rule was a vital piece of equipment; carefully transferring measurements from it using special beam compasses, a twenty-foot master scale was marked out. This scale was next used to set a twenty-foot beam compass, which was used to mark out the instruments. These included a steel chain one hundred feet in length (also in the Science Museum's collections) and three twenty-foot long wooden rods with a standard metal rod for comparison. Measurements would be made with both to ensure that the results would be more accurate. The chain proved to be even more accurate than had been hoped, and remarkably quick and easy to use. Two different methods of measurement were attempted with the wooden rods, one of which (putting the rods end to end) was found to be very much more accurate than the other. Although wooden rods had been used to measure most of the other base lines in other countries, this method proved unsatisfactory, as humidity changes caused large variations in the lengths of the rods in the course of each day. When it became clear that this was the case, Roy took up a suggestion from Lieutenant Colonel Calderwood that glass tubing might be used instead, and a week of rainy weather was devoted to the construction of a suitable number of glass tubes and converting them to measuring rods. The twenty-foot master scale was rechecked using the Sisson rule and then used to check the length of the glass rods.

In the end the base was measured using both glass rods and the 100-foot chain. Work started again in August, and on Saturday 21 August King George III came out to Hounslow Heath to see how it was progressing. By the end of August the base line had crossed thirteen roads and two rivers, and was finally calculated to be 27404.7219 feet in length.   JEI

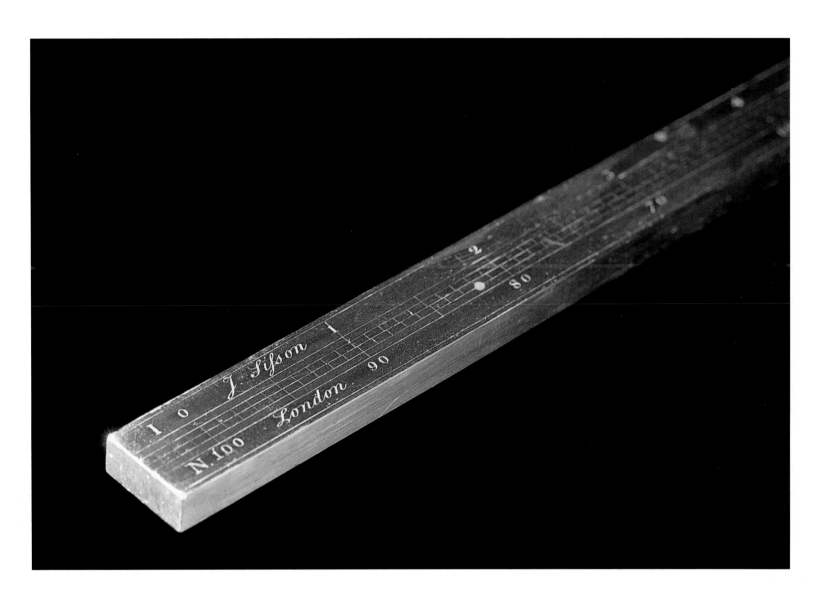

SHELTON'S ASTRONOMICAL REGULATOR

In the 1760s John Shelton (fl. 1712–77), clockmaker in Shoe Lane, London, supplied five astronomical regulators to the Royal Society and the Board of Longitude. These were specifically intended for timing the transits of Venus which took place on 6 June 1761 and 3 June 1769, but they had a long and remarkable subsequent career. All five still exist.

An astronomical regulator is simply a very exact clock used for timing astronomical events. The most important components are the escapement and the pendulum, and Shelton made use of two of the innovations of eighteenth-century horology. The 'dead-beat escapement' had been pioneered by his former employer, George Graham (*See* THE ORIGINAL ORRERY), in about 1720, and it was so accurate that it continued to be used in precision clocks for another two centuries. The 'gridiron pendulum' invented by chronometer builder John Harrison (1693–1776) counteracts the effects on the clock of changes in temperature – particularly important for these clocks, which were afterwards to travel from the Arctic to the Tropics.

The clock features a typical 'regulator dial'; the outer dial measures the seconds, the smaller dial inside measures the minutes, and the dial segment seen beneath the centre measures the hours. This was a convenient arrangement for astronomers, who had to identify at a glance the exact second of an observation. The wooden case is not the original; it is a replacement made in 1884 in Washington, DC, at the end of the clock's final expedition.

The pair of transits of Venus in the 1760s roused immense interest all over Europe; this rare phenomenon, when the planet Venus is seen passing across the Sun's disc, could be used to measure the distance of the Sun from the Earth – if only it could

be accurately observed at widely separated places. In one of the earliest examples of international scientific cooperation learned institutions rose to the challenge and mounted many programmes of observation.

All five of the Shelton regulators took part in one or other of the British transit expeditions, one of them being used for observations as far away as Tahiti, during Captain James Cook's first voyage (1768–71). It is not now possible to distinguish the early history of each instrument, but it appears that the regulator in the Science Museum was one of those that accompanied the transit expedition in 1769 to the far north of Norway (Captain William Bayly to the North Cape, and Captain Jeremiah Dixon to Hammerfest). A year or two later it seems to have been one of the two precision clocks taken by Captain Cook on his second and third voyages of discovery to the South Seas (1772–5 and 1776–80).

One of the five Shelton regulators was used in one of the most famous experiments of the eighteenth century, Nevil Maskelyne's measurement in 1774 of the weight of the Earth. He used it to see how much a pendulum's rate was influenced by a neighbouring mass – Mount Schiehallion in Scotland. Many years later, in 1828, the regulator pictured here was used by the Astronomer Royal, George Airy, in another experiment to weigh the Earth – comparing the rate of a pendulum at the top and the bottom of a deep mine, Dolcoath copper mine in Cornwall.

However the Shelton regulators found their fullest use in yet another important astronomical problem, measuring the exact shape of the Earth. The rate of a pendulum depends on how far it is from the centre of the Earth. It will swing more slowly at the top of a high mountain than at sea level, so if enough

measurements are made worldwide then the shape of the Earth can be established. The Shelton regulators were used for these 'pendulum' measurements as early as the 1760s, and they later accompanied many expeditions all over the world.

This regulator was with the British scientist Edward Sabine on four separate pendulum expeditions in the North and South Atlantic, (1818–23) reaching as far north as the 79th parallel in Spitzbergen. In 1865–73 it travelled the length and breadth of India with Captains J P Basevi and W J Heaviside of the Great Trigonometrical Survey of India. In 1882–4, on loan to the United States Coast and Geodetic Survey, it saw service in Australasia, Singapore, Japan and the United States. In Auckland the Americans used it for the transit of Venus which took place in 1882, 113 years after the famous transit of 1769 for which it had originally been constructed.

Not much is known about Shelton's life. He was apprenticed in 1712, and later he worked for George Graham, one of the most distinguished of eighteenth-century clockmakers; and in particular he was Graham's chief constructor of astronomical clocks. But in 1777, then in his eighties, he was destitute and presented a petition to the Royal Society 'setting forth that a great many clocks and astronomical regulators have been made by him ... which are in position in the chief observatories of the world ...'. The outcome of this appeal is not known.

The art of precision clockmaking was well developed by the second half of the eighteenth century, and Shelton's regulators do not present any significant innovations of his own. But they are remarkable for their long life; there can be few scientific instruments which have travelled so far over such a long period for such a variety of uses. GF

ARKWRIGHT'S SPINNING MACHINE

Perhaps the most visible and dramatic memorial of industrial capitalism as it was pioneered in Britain in the second half of the eighteenth century has been the factory. Here was a new and specialized type of building designed not only to make efficient use of new forms of machine, to which water or steam power could be applied, but also as a setting for the direction and management of labour – the factory system.

The factory, or mill as it was known in the textile industry, became a potent symbol of the triumphant power of industrial production and the oppression and degradation of workers. It became the scene of violent machine-breaking by those who saw their livelihoods threatened, the focal point of the new industrial towns and the sole source of their great wealth and prosperity. Paradoxically in recent years it has been the closure of factories that has been blamed not just for the extinction of jobs but for the destruction of the social spirit which many of those industrial communities cherished.

The first factories, in the form that we have come to recognize them, were built to house new machines for spinning yarn. They were introduced into an industry which had hitherto been conducted on a domestic scale. Arkwright's machine and the factory organization he introduced to exploit it changed all that, suddenly and dramatically. That change in preparing fibres and spinning the yarn soon spread to other branches of the textile industry and in turn that industry set the pattern for others.

Sir Richard Arkwright (1732–92) had little or no schooling and began his career as a barber. Later he travelled, collecting hair for wig-making, and that may have given him the opportunity to hear of the numerous efforts then being made to improve production in the textile industry – attempts on which he was to base his own ideas. He was probably not much of an original inventor, but he was exceptionally good at seizing on a promising idea and exploiting it.

Arkwright's forceful approach made him many enemies, but he became a wealthy man. He was knighted in 1786 and was appointed High Sheriff of Derbyshire in 1787. His portrait, by Joseph Wright of Derby, also to be found in the Science Museum, makes him appear complacent, arrogant and powerful, not an engaging man. Matthew Boulton, a contemporary industrialist of a very different stamp wrote of him: 'Tyranny and an improper use of power will not do in this country . . . If Arkwright had been a more civilised being and understood mankind better, he would now have enjoyed his patent'.

In spinning, fibres are drawn out – drafted – and then twisted together. Arkwright's machine uses the 'flyer' of the conventional hand spinning wheel to provide the twist. But whereas the hand spinner drafts the fibres by teasing them down from the distaff, in Arkwright's machine the action of the fingers is simulated by pairs of 'drafting rollers'. There are two pairs of these to each thread. The second set are driven slightly faster than the first so

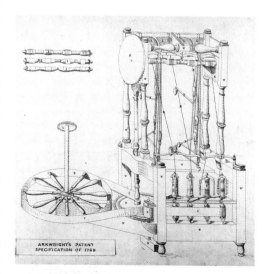

ARKWRIGHT'S PATENT
SPECIFICATION OF 1769

The published lithograph of a drawing accompanying Arkwright's patent specification of 1769. The drawing is based on the Science Museum's machine, showing how it was driven and other details now lost

that the fibres are teased out between them.

This use of rollers was not new. It may be traced back to Lewis Paul's patent of 1738. But Arkwright's machine was better arranged. The small machine in the Science Museum corresponds so closely to the drawing of Arkwright's patent of 1769 that it is likely to be his prototype or development model from that time. It bears the scars of having been modified, as one would expect. This machine originally had four spindles and might have been suitable for use in the home, according to the old domestic system of organization. In fact, this was not done; Arkwright moved directly to the grouping of many spindles together under one roof, to be driven by a central source of power.

His first mill, at Nottingham, was driven by a horse. It has long since disappeared. But in 1771 Arkwright and his partner built a larger mill, worked by water power. This was at Cromford in Derbyshire and it still survives, although now robbed of its top storey. Arkwright sold licences to a few other spinners permitting them to use his machine, but it seems certain that the water frame, as it came to be called, was widely pirated in the 1780s. There are many remains of early spinning mills, especially in the north Midlands, and their remarkable conformity to the standard width of approximately thirty feet set by Arkwright's own factories suggests that they were built to house water frames. After 1790, when mule spinning was gradually adopted, new cotton mills were built fifteen or more feet wider to accommodate the new machines.

Arkwright licensed his innovation only in units of a thousand spindles, thus forcing licensees to adopt a system similar to his own. He seems to have done this primarily in an attempt to control the use of his patents; but the effect of encouraging the grouping of large numbers of machines together, sharing a central power source and with the organization and division of labour to attend them, was to determine the pattern of development of the factory system in the nineteenth century. NC

TROUGHTON'S DIVIDING ENGINE

In the eighteenth century, the scales on scientific instruments were marked out by skilled craftsmen in a process called 'dividing', hence accurate instruments took a long time to make and were expensive. From the 1770s dividing was partly mechanized as 'dividing engines' were introduced, enabling instrument makers to satisfy a growing demand for improved navigational instruments. At first, they were only capable of dividing small instruments, but by the 1850s dividing engines were being used to graduate the scales on large astronomical telescopes, and hand-dividing was obsolete.

This dividing engine was completed by John Troughton (c 1739–1807) in 1778, and is important because it is similar to the first successful example, completed by Jesse Ramsden (1735–1800) about 1775.

The incentive to develop these tools came from a major navigational problem: how to find a ship's longitude. In the eighteenth century Britain depended on its navy for defence and its merchant ships for trade, so accurate navigation was vital. The British Government set up the Board of Longitude in 1714, which offered a top prize of £20,000 (equivalent to at least a million pounds today) for a way of finding longitude to within thirty nautical miles; and there were smaller awards for related innovations.

By the 1760s John Harrison and Tobias Mayer had developed practical methods of finding longitude at sea; the former by building an accurate chronometer, and the latter by tabulating the motion of the Moon. Both men were rewarded by the Board. Their methods needed accurate angle-measuring instruments, such as the octant and sextant, to measure the altitudes of heavenly bodies, and spurred a demand for these instruments, but accurately dividing such small scales by hand slowed production.

The English instrument maker Jesse Ramsden provided the solution, by mechanizing dividing. Clockmakers had used machines to position and cut gear teeth since the late seventeenth century, and about 1740 a York clockmaker, Henry Hindley, built a

Ramsden's second dividing engine; engraving from *Description of an Engine for Dividing Mathematical Instruments* by Jesse Ramsden, 1777

more accurate version which could also be used to divide instrument scales. Hindley's machine was the earliest dividing engine and incorporated the worm-and-gearwheel arrangement used in later engines. It was used to divide a few astronomical instruments, but remained little known. However, one of his workmen, John Stancliffe, later became Ramsden's foreman, and may have told him of Hindley's machine. Ramsden's first dividing engine was completed about 1766, but was not accurate enough for navigational and astronomical instruments. His second engine, completed about 1775, was a complete success. The Board of Longitude appreciated its importance and wanted to make it available to all instrument makers. Ramsden was paid £300 for

disclosing its secrets, and the Board bought the engine for £315 while allowing Ramsden to retain and use it. As a condition of the payment, between October 1775 and October 1777 Ramsden was to train up to ten other instrument makers how to duplicate and use the engine. He also had to publish a description of the engine, with enough detail to allow other instrument makers to duplicate it.

John Troughton was probably one of the instrument makers taught by Ramsden. He had a high reputation as a hand-divider and was well placed to appreciate the importance of this new tool. In 1778 he completed his dividing engine, which had taken him some three years of spare time work to build. As Ramsden's book on the dividing engine was only published in 1777, it is difficult to see how else John Troughton could have got full details of Ramsden's work.

Troughton's engine was profitable; Ramsden had a reputation for long delivery times and occasional bad workmanship. Instrument makers preferred to pay more for a quicker dividing service from John Troughton, who was said to have made more money with his engine than Ramsden got from the Board of Longitude. It earned him enough to purchase the instrument making shop of Benjamin Cole at 136 Fleet Street, London, in 1782, later to form part of the famous firm of Cooke, Troughton and Simms.

The dividing engine was simple to operate. The instrument being divided was fixed to a large wheel on top of the engine. When the treadle was pressed, the wheel and the instrument were turned through a fixed angle. Then with the right hand, a cutting tool guided by a system of swinging links was used to mark the instrument scale. The process was repeated until the complete scale had been divided. Although this was many times faster than hand-dividing, it was a backbreaking task to lean over the engine to work on a small sextant. John's brother Edward wrote that 'it had done no good either to his health or that of my own, and had materially injured that of a worthy young man then my assistant'. PJT

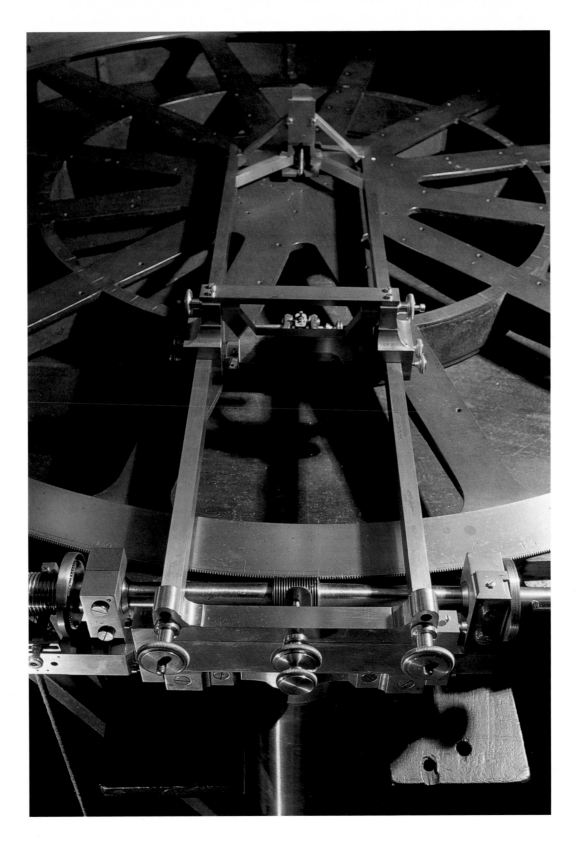

HERSCHEL'S SEVEN-FOOT TELESCOPE

During the eighteenth century, few people had a greater impact on astronomy than William Herschel (1738–1822). At a time when astronomers were preoccupied with measuring the position of stars to ever greater levels of precision he was busy exploring the structure of the universe. He was also active in advancing telescope technology. Herschel's serious interest in astronomy did not develop until he moved from Hanover to England in 1773. As a professional musician and composer he had gained the post of Director of Music for Bath. In his spare time he developed his interest in astronomy and having found commercial telescopes unsatisfactory he learnt how to make his own. Herschel built his telescopes to a design first suggested by Isaac Newton, which used mirrors instead of lenses to focus light. With time he perfected his instruments until they were superior to any others then available. This led to his discovery of Uranus in 1781 with a telescope similar to the one shown here. The discovery made Herschel famous overnight, as no new planets had been observed since antiquity. He originally named the planet 'Georgium Sidus' (George's Star) after the reigning King George III, but fellow astronomers did not agree with his choice. In 1782, Herschel was appointed 'Royal Astronomer' and this royal patronage allowed him to give up his musical duties and move to Datchet near Windsor where he could devote himself fully to astronomy.

Herschel made and sold many telescopes during his long career. This example was made for Caroline Herschel (1750–1848), his sister, and was a present from him. Caroline was a very capable astronomer in her own right and acted as William's assistant when he was observing. Her contributions to astronomy have been somewhat overshadowed by those of her more famous brother. She is best known for her careful observations of comets; between 1786 and 1797 she discovered a total of eight comets. This was a remarkable achievement for the time and is still impressive even today. She also undertook the onerous task of correcting the numerous errors in Flamsteed's star catalogue, *Historica Coelestis*, the standard work of its day which William used in his searches of the heavens.

Silhouette of Caroline Herschel c 1772;
the only known portrait of her as a young woman
(Museum of the History of Science, Oxford)

Caroline stated that this telescope was a copy of the one that William used to discover Uranus. It had a six-inch-diameter mirror with a focal length of seven feet. (The image is formed seven feet from the mirror.) The mirror was made of speculum metal, a highly reflective but very hard alloy of copper and tin which can only be ground or polished with difficulty. Unfortunately, speculum mirrors readily tarnish and therefore need to be repolished frequently. The design of this telescope is typical of instruments made through Herschel's career, and had an influence on telescope makers for over a hundred years. Astronomers needed larger telescopes to see fainter objects and to distinguish more detail. Herschel showed how to make them at a cost modest compared to other designs then available (*See* THE ROSSE MIRROR).

Using the experience that he had gained with his smaller telescopes Herschel was able to construct two instruments with a focal length of twenty feet. One was built whilst he was still living in Bath, but it proved difficult to use due to the small size of his garden. With his new instrument Herschel embarked on a systematic survey of the heavens. This was to be the most productive period of his life. He was able to confirm his earlier observations whilst making new discoveries. He was also able to uncover many more double stars, some of which he proved were rotating around each other. There had been a long debate during the eighteenth century as to why these stars appeared double. Some astronomers thought they were simply 'a line of sight effect' and that the stars lay in the same direction but were at different distances. Herschel's observations over many years proved that some moved in orbit around each other. These successes spurred Herschel to build an even larger telescope and, with patronage from the King, he was able to cast, polish and grind several speculum mirrors of forty-eight inches in diameter. The resulting forty-foot telescope was erected at Herschel's new home in Slough in 1789 where it became an instant landmark due to its sheer size. During its construction the King was a frequent visitor and delighted in conducting people through the enormous telescope tube. On one occasion whilst with the Archbishop of Canterbury, the King was heard to say, 'Come, my Lord Bishop, I will show you the way to Heaven!'. Alas, the telescope was not as successful as hoped and was little used. This was partly due to its huge bulk which took three to four people to operate. The instrument fell into disuse in 1815 and was finally dismantled in 1839.

On William's death in 1822 Caroline Herschel returned to Hanover where she lived until her death at the age of ninety-seven. Such was her fame in her later years that she was awarded a gold medal by the Royal Astronomical Society. In addition, she was made an honorary member of the Royal Society, the first woman to receive this accolade. KLJ

BOULTON AND WATT ROTATIVE ENGINE

'I sell here, Sir, what all the world desires to have, power!' was the proud boast of Matthew Boulton (1728–1809) to a visitor attending the famous Soho Manufactory in Birmingham. What Boulton was selling, with his partner James Watt (1736–1819), was the power of steam, the single most important technological factor in bringing about the huge economic and social changes that have been called the Industrial Revolution. Steam drained mines, powered machinery, propelled locomotives and ships and had helped transform Britain and many other countries into highly industrialized societies. Even today, although the older reciprocating engine has been largely replaced, it is still steam power, in the form of turbine driven alternators producing electricity, that satisfies a high proportion of the nation's energy needs.

James Watt did not invent the steam engine, nor did his engines lead to the emergence of the factory system, which was pioneered initially on water power. What he did contribute however were outstanding improvements in efficiency and an engine that would progressively supplant the water wheel, thus freeing factory owners from the geographical and seasonal limitations of steeply graded river valleys.

Thomas Newcomen (1663–1729) had built the first recorded engine in 1712 and the machine was well established for mine drainage before Watt was born. As a young instrument maker, Watt was asked to improve a model of Newcomen's engine intended to demonstrate the working principle. His appreciation of its inherent inefficiency – the thermal losses resulting from raising and lowering the temperature of the working cylinder during the cycle from steam inlet to condensation – led him to a brilliant solution, the separate condenser, the greatest single improvement ever made to the steam engine. By removing the location of the condensation from the cylinder to a separate evacuated chamber, so that the temperature of the working cylinder was not significantly lowered, he brought about a major increase in thermal efficiency and absolute power.

Watt patented his condenser in 1769. In 1782, he introduced the double-acting engine in which steam was applied alternately to both sides of the piston, thereby obtaining double the power from the same size of cylinder. In the same year he took out a patent for a rotative beam-engine in which the beam drove a shaft and flywheel. These features, together with his parallel motion, which provided a positive linkage between the reciprocating end of the piston rod and the beam meant that Boulton and Watt had an engine capable of driving machinery. By 1800 when Watt's partnership with Boulton ended and the patent on the separate condenser expired, 451 engines had been built of which 268 were rotative.

The engine in the Science Museum was built in 1788 to power a section of Boulton's own works and incorporates all the features developed to that date. The connecting rod is linked to the flywheel shaft through sun-and-planet gearing, possibly devised to avoid infringing James Pickard's patent for the common crank. It is also the first engine ever fitted with a centrifugal governor to regulate its speed.

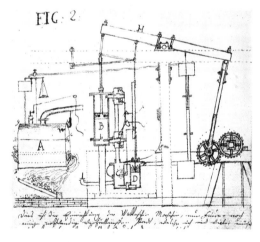

A sketch, probably based on this engine, made by the German mechanic Georg Reichenbach in 1791. Some details, since altered, are shown in their original form, whilst some internal details have been guessed at and are incorrect

The engine proved the soundness of its design and construction by giving seventy years of good service. As with any machine that has been so long in use there have been repairs and the replacement of parts, but it is believed that this is the oldest engine in the world that is substantially complete and in its original state. At this time Boulton and Watt still did not in general build whole engines. Broadly, they followed the earlier practice of providing a design and advising over the sources of supply of major parts or materials. They would offer a trained erector to oversee the building of the engine and usually provided some critical components such as the 'nozzles' or valve chests. Therefore no two engines were quite alike.

The engine also seems to have been the object of industrial espionage. In 1791 the young German mechanic Georg Reichenbach noted in his diary how he bribed the night watchman to let him make repeated visits to sketch it. Returning to Germany, Reichenbach was responsible for the design of similar engines.

The 1788 engine marks not only the introduction of steam power to a wide range of industrial applications but reveals a number of general truths about innovation and technology transfer. Boulton and Watt dominated the market after 1780 but it was in part the result of their energetic defence of their patents. Watt became progressively more conservative, set his face against high-pressure steam engines and was reluctant to consider steam for road or rail traction or engines for boats. Although by 1800 Watt's engines were still the most reliable, largely as the result of superior design and workmanship, they were also by then old-fashioned. Scores of other engineers were pioneering new developments. By the end of the eighteenth century perhaps 1,200 to 1,330 steam engines had been built, nearly 500 of these to Watt's designs. Before the end of the nineteenth century the steam engine had become the universal source of power for industry and transport.   NC

SYMINGTON'S MARINE ENGINE

Mechanical propulsion of ships is one of the topics which has fascinated experimenters ever since the Renaissance. They would all have been thoroughly familiar with the waterwheel, the origin of which is known to predate the Roman era. The waterwheel was thus an obvious prototype to be copied in trials of mechanical propulsion of ships. Among the many eighteenth-century figures attracted by this quest was a wealthy Edinburgh banker, Patrick Miller (1731–1815), who commissioned a sequence of double- and triple-hulled vessels with paddle wheels between their hulls which were turned by manual capstans. However, human stamina proved insufficient to propel any of these boats for more than a few minutes. Miller was about to discontinue his experiments when he learned from the tutor engaged to educate his sons about a 20-year-old engineer named William Symington (1763–1831) working in the Lanarkshire lead mines. Symington had successfully built an experimental beam engine combining the thermal efficiency of a separate condenser with the simplicity of an atmospheric engine on Newcomen principles. In 1786 Patrick Miller witnessed a demonstration of a steam carriage fitted with one of Symington's engines. When Symington was granted a patent the following year, Miller was sufficiently confident of the engineer's potential to order an engine to power a catamaran on the lake at his Dalswinton estate in Dumfriesshire.

Symington's response was a wooden-framed atmospheric engine, whose power came from the vacuum created when a cylinder full of steam is suddenly condensed. Because each of the two vertical pistons created power only on the down stroke, they were linked by chains to a drum suspended above them, which revolved in alternate directions as the pistons rose and fell. The machine's principal ingenuity lay in Symington's use of ratchets and pawls (previously patented by Matthew Wasbrough, in England only, in 1779) to rectify semicircular motion into continuous rotation. Chains connected the working drum with a pair of loose pulleys on each of the two paddle shafts. Ratchet teeth around the interior of each pulley engaged alternately with opposite facing pawls keyed to the two paddle shafts. Interconnection between the sets of pulleys ensured that both paddle shafts rotated continuously in a uniform direction as the pistons rose and fell in their open-topped brass cylinders. This ingenious alternative to the crank previously patented by James Pickard is claimed to have propelled Miller's steamboat at five knots in trials on Dalswinton Loch. Like many other claims advanced by early steam engine promoters, this seems highly exaggerated.

Whatever the vessel's performance, Miller was sufficiently encouraged to commission a much larger engine of almost eleven horsepower from his protégé, and Symington's next engine successfully propelled a sixty-foot catamaran on the Forth and Clyde Canal near Falkirk in December 1789. But at this larger scale the ratchet arrangement revealed its inadequacy for sustained power transmission under

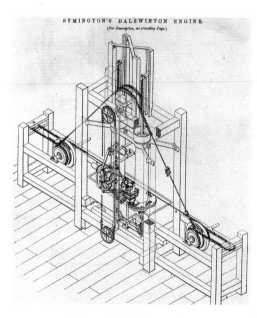

Symington's Dalswinton engine,
from *Engineering*, 26 January 1877

load, and the pawls failed repeatedly. A curt rebuff from James Watt to Miller's proposal for a collaborative development prompted the Edinburgh banker to give up steam engineering and devote his energies to agricultural improvement; Miller was also apprehensive of litigation by Boulton and Watt over alleged infringements of the separate condenser patent. He removed Symington's smaller marine engine to his library where it remained as a curio. This was not the end of Symington's involvement with steamboats however. In 1801 he produced a different type of engine for the experimental canal tug *Charlotte Dundas*; however the Forth & Clyde Canal Company decided not to use steam tugs as their wash would damage the canal banks.

Miller's son inherited the engine on his father's death in 1815. After years of neglect under its new ownership, the decision was made to sell it for scrap. An Edinburgh plumber got as far as discarding the worthless wooden framework and separating the non-ferrous components when he died. In 1853, agents for Bennet Woodcroft (1803–1879), himself a pioneer of steam navigation, located the surviving components in the plumber's workshop and purchased them for the Patent Museum.

To make a meaningful exhibit of these, extensive reconstruction and replacement were necessary. Woodcroft entrusted this work to John Penn's engine works at Greenwich. A new wooden framework had to be constructed, although this was difficult since there were no detailed drawings in Symington's patent specification. Some of the valve gear and all of the chains and ratchet mechanism had to be replicated. So thorough was the rebuild that the engine could be worked in steam.

Predating as it did Henry Bell's *Comet* steamship (*See* THE 'COMET' STEAM ENGINE) by more than twenty years, the engine conceived by William Symington to advance the progress of steam power in the face of antagonism from James Watt and his colleagues deserves its revered status as the oldest steamship engine surviving anywhere in the world.   JCR

RAMSDEN'S THREE-FOOT THEODOLITE

It is occasionally the case that the finest peak of achievement in one area of science or engineering becomes the starting point for a new, quite different one. This three-foot theodolite of 1791, by Jesse Ramsden (1735–1800), is an example.

The theodolite remained in constant use for over sixty years. It was used for measuring the angles in the Primary Triangulation of Great Britain, the foundation for the first Ordnance Survey maps of the country. This huge undertaking was initiated by General William Roy (*See* SISSON'S RULE), with a trigonometrical operation to fix the relative positions of the observatories of Paris and London more accurately than ever before; indeed so accurately it would be 150 years before the measurements would be done again. The first stage was the measurement of the length of a base line on Hounslow Heath in 1784. One end, still marked by a cannon, is a National Monument lying within the boundary of Heathrow Airport.

The survey was to reach across England towards Dover; then sightings across the channel would link the English and French surveys. The scheme, suggested by Cassini de Thury through the French Ambassador to King George III, was handed to the Royal Society to implement, with royal funding. Sir Joseph Banks (1743–1820), the President of the Royal Society, knew that William Roy had argued persuasively for over thirty years to set up an accurate mapping framework nationwide, and promptly asked him to undertake the work. Roy leapt at the challenge, as the project could well develop into the national triangulation he had long advocated, as long as the work was shown to be of the highest quality.

Jesse Ramsden was one of the top ten specialist instrument makers in eighteenth-century England, supplying large-scale astronomical and surveying instruments both at home and abroad. He became a Fellow of the Royal Society in 1786. Ramsden's first three-foot theodolite was made for the Royal Society

and took three years to complete. When it was finished in 1787, Roy could take a bearing on a mark at up to 70 miles away with an error in the angle reading of only one 1/180th of a degree, and was in a position to measure the curvature of the Earth.

The second instrument made by Ramsden – and now in the Science Museum – was even better and

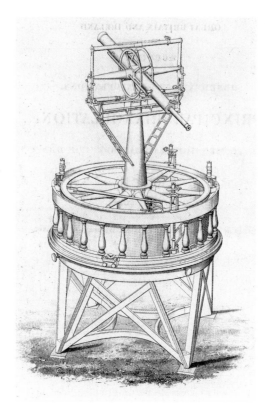

Ramsden's Great Theodolite:
frontispiece to *An Account of the Observation and Calculations of the Principal Triangulation*, 1858.
The horizontal circle, carrying the scale, was the most important part of the instrument, requiring to be kept as flat and stiff as possible. The instrument had to withstand an immense amount of wear and tear in constant use, rough travelling and exposure to the weather

the accuracy of its scales was only exceeded in the 1930s. It was commissioned by the Survey of India, but Ramsden incorporated improvements to the design which increased the cost to £373, 14s, beyond what the Survey of India was prepared to pay. Fortunately for Ramsden, the Third Duke of Richmond, Master General of the Ordnance, agreed to purchase his white elephant, and the Principal Triangulation of Great Britain was formally authorized with the sanction of King George III.

Both theodolites were used for the most important of the 229 stations in the network, and despite their enormous weight were transported to the tops of mountains, steeples or even scaffolds specially built to take them, so the lines of sight could be maintained.

In installing the theodolite before taking a measurement, the first problem was to provide a perfectly solid foundation. This occasionally required digging; in Holme Moss, nine feet of bog and six of sand had to be removed to get down to a steady base. On exposed mountain tops, a wall of stone would be built around the base of the observatory tower. This precaution certainly saved the instrument one day in 1838, when a storm moved the temporary observatory out of position, even though the protective wall was two feet thick.

The names of the observers and the 'bookers' (the men who wrote down the readings in the observation books) were recorded at every station. All the calculations and subsequent map drawings were carried out at the office. For the first fifty years of the Ordnance Survey this was at the Tower of London. After a disastrous fire in 1841 in a building next to the Ordnance Map Office, the Survey moved to Southampton. Nearly 100 years later, the first Ramsden theodolite fell victim to a World War Two air attack during which the building was almost completely destroyed, so the Board of Ordnance theodolite is the oldest of its kind to survive. JEI

MAUDSLAY'S SCREW-CUTTING LATHE

The successful construction of all machinery depends on the perfection of the tools employed ... The contrivance and construction of tools must therefore ever stand at the head of the industrial arts.
Charles Babbage, 1851

Henry Maudslay (1771–1831) began work at twelve years of age filling cartridges at Woolwich Arsenal, eventually becoming a highly skilled blacksmith. At eighteen he was recommended to Joseph Bramah as the man to solve problems in the manufacture of Bramah's patent lock, and he worked for him until 1797 when he set up on his own. From the outset Maudslay always strove to work 'in the best possible manner', and this lathe designed specifically for making screws reflects his commitment to perfection.

Screws are vital components in machinery. They are used not only as fastenings but also as a means of adjustment; for these purposes no great precision is usually demanded but it is convenient to have them well fitted, smooth in action and preferably interchangeable one with another. But a more demanding role is their use for measurement and to give regular, controlled motion in machinery, and for these purposes a much higher standard of precision is needed. But screws were usually made by very crude methods, so that these properties could not be taken for granted.

The modern method of making a screw with any pretension to accuracy is to rotate a cylindrical workpiece in the lathe while a cutting tool held by a slide rest is automatically moved along to form a helical groove in the workpiece. Usually the slide rest is moved by a leadscrew geared to the rotating workpiece. Maudslay was not the first to use either the slide rest or the leadscrew, both of which are essential parts of this machine. It is arguable that

Henry Maudslay: lithograph by H Grevedon, 1827. Photographed from the original portrait on stone in the collections of the Science Museum

Maudslay was unaware of any precedents, but in any case his remarkable achievement was to have introduced such tools into everyday use.

The lathe is built on two parallel triangular bars. One bar carries conventional headstocks to hold the workpiece. A slide rest riding on both bars carries a tool holder which is equipped with a screw feed having a micrometer dial to regulate the depth of cut. The leadscrew carried between the bars moves the slide rest along; it is so mounted that it may readily be changed. The gears to connect the leadscrew to the mandrel rotating the workpiece are lost, as is the means of driving the machine, a capstan wheel worked by hand.

In 1846 the famous lathe manufacturer and machinist Charles Holtzapffel (1806–47) gave a detailed description of how it was used. The problem that faced Maudslay was how to originate an accurate screw thread without copying the errors in a pre-existing master. His technique was to originate a new screw thread by using a tool with an inclined knife which cut a helical trace on a soft metal rod. Selected screws were used as leadscrews in this lathe to cut new screws in any material. Various techniques were applied to smooth and average errors, to arrive at ever more perfect screws.

Using this and similar machines Maudslay went on to produce tools for the production of ordinary screws as well as screws for all types of machine, including those of the utmost precision for measurement. In his works he introduced a bench comparator, capable of measuring lengths to a tenth of a thousandth of an inch. It was known as 'The Lord Chancellor' as it was regarded as the final court of appeal in matters of accuracy.

It was a particularly well made screw that Maudslay had displayed in the window of his first little workshop that gained him the contract to build Brunel's block-making machinery (See PORTSMOUTH BLOCK-MAKING MACHINERY). An amateur mechanic, M. de Bacquancourt, admiring the screw, went in to find out how it had been made; the visit made a lasting impression. When de Bacquancourt heard that his acquaintance Marc Brunel needed a good workman to realize his mechanical ideas, he effected the introduction that made Maudslay's career.　　MTW

During the period from 1800 to 1810, Mr. Maudslay effected nearly the entire change from the old, imperfect, and accidental practice of screw-making ... to the modern, exact and systematic mode now generally followed by engineers.
Charles Holtzapffel, 1846

HERSCHEL'S PRISM AND MIRROR

At first glance these objects look insignificant. One is a small prism of flint glass about five inches long, the other an ellipsoidal metal mirror attached to a square frame. Their significance lies in the fact that they were used by William Herschel (See HERSCHEL'S SEVEN-FOOT TELESCOPE) in the first investigations of the spectrum beyond the visible region, now known as the 'infra-red'. This pioneering work is probably less well known today than Herschel's discovery of the planet Uranus or his construction of large telescopes, but the implications of extending the spectrum were far-reaching.

In 1799, during the course of his research on sunspots, Herschel on many occasions viewed the Sun through a telescope fitted with protective lenses. To cut out the glare he used densely coloured glasses in the eyepiece. However this was beset by problems; if the glare was satisfactorily reduced an intolerable heat was transmitted, while if the heat was reduced the lens transmitted too much light. Sometimes the glass cracked, on one occasion 'with a very disagreeable explosion that endangered the eye'. Having tried many colours he was led to question whether, in his words, 'the power of heating and illuminating objects might not be equally distributed among the variously coloured rays.'

The first experiments, published by the Royal Society in 1800, compared the heating and lighting effects of the various colours. Using this small prism Herschel made a spectrum of sunlight fall on to a table where he then placed three thermometers. He used a piece of card with a hole in it to select each colour in turn, the colours being those defined by Newton: red, orange, yellow, green, blue, indigo and violet. One thermometer was placed in the coloured ray while the others were used as controls, simply measuring the temperature of the room. He found that the red gave considerably more heat than the others, the quantity decreasing to almost nothing for the violet ray. He then used the prism in a second experiment to view objects under a microscope in different coloured light. Judging by his eyes he found green/yellow to be the best for illumination.

Herschel found that dark green smoked glass gave the best results for his telescopic researches, but more interestingly he showed that the maximum heating effect and lighting effect are at different points on the spectrum. A more tantalizing prospect was that the maximum heating effect might lie 'even a little beyond' the red end of the visible range. Herschel set out to investigate this possibility.

Again the prism was used to form a spectrum but this time one of the thermometers was placed beyond the red end. A card was marked in inches to enable the position of the maximum heat to be found. As he expected, Herschel found the greatest heating effect just beyond the red but no effect beyond the violet. This was the first time that experiments had been made on the spectrum outside the visible region.

Herschel's apparatus for determining the position of maximum heat

Herschel went on to perform many experiments on 'the rays which occasion heat'. He used his kitchen fire, a candle flame and a red-hot poker. He showed that heat could be reflected, refracted or bent, scattered and transmitted in much the same way as light. The mirror was used for experiments on the solar spectrum. The square frame holding the mirror was attached to a wall or shutter and the mirror turned slowly to keep the Sun's rays reflected on to the prism.

Likely nearly all of his contemporaries Herschel believed that light consisted of a stream of particles. In the description of the first experiment he surmised that 'radiant heat consists of particles of light of a certain momenta'. Paradoxically, in the same year Thomas Young (1773–1829) proposed his wave theory of light. Although this was not well received, Herschel had to take account of Young's work and referred later in the year to 'vibrations which occasion heat or rays'.

Although an explanation of Herschel's findings based on the wave theory of light had to wait until the 1830s, there was a faster response to the idea of infra-red from Johann Wilhelm Ritter (1776–1810) of Silesia (now Poland). Believing in the harmony of nature he looked for an effect at the violet end of the visible spectrum. In 1801 he found that these rays turned paper soaked in silver chloride solution black, and hence was the first to observe 'ultraviolet' radiation.

We now believe that the 'infra-red', 'visible' and 'ultraviolet' regions are a small part of an entire electromagnetic spectrum which includes radiowaves, microwaves, X-rays and gamma rays. All these types of radiation are simultaneously particles and waves, and travel at the speed of light. Obviously a great deal of thought and experiment has been put into this area of physics since Herschel's day, yet he took the first step with this simple apparatus.   JAW

'COALBROOKDALE BY NIGHT'

. . . if an atheist who never heard of Coalbrookdale, could be transported there in a dream, and left to awake at the mouth of one of those furnaces, surrounded on all sides by such a number of infernal objects, though he had been all his life the most profligate unbeliever that ever added blasphemy to incredulity, he would infallibly tremble at the last judgement that in imagination would appear to him.

With these apocalyptic words Charles Dibdin, dramatist and song-writer, concluded his observations on the Severn Gorge in Shropshire. What drew Dibdin and numerous other travellers here in the late eighteenth and early nineteenth centuries was the remarkable concentration of industrial activity, concerned in the main with ironmaking, spread along some three miles of the valley of the River Severn between Coalbrookdale and Coalport.

The pioneering work of the Quaker ironmaster Abraham Darby in developing coke smelting in 1709 had been followed by a sequence of new uses for iron – the first iron rails, the first iron bridge, iron boat and steam railway locomotive – making Coalbrookdale an area of interest irresistible to anyone concerned to understand the new forces of industry that were sweeping Britain. Thomas Telford, William Jessop, Josiah Wedgwood, Richard Trevithick, James Watt, Matthew Boulton and John Loudon McAdam were the most notable of the British engineers and entrepreneurs who came to Coalbrookdale. They were joined by visitors from France, Prussia, Bavaria, Sweden and North America, many of whom were to leave accounts of what they saw. At a time when almost all tourists kept a journal of their travels, and many published them, these reflections provide a vivid and often technically detailed record of an area that the Shrewsbury cotton master, Charles Hulbert, was to describe in 1837 as '. . . the most extraordinary district in the world'.

The gorge exercised a peculiar fascination for artists too. Here the combination of topography – the River Severn flowing through its precipitous gorge – and the smoke and flames of the furnaces and forges conspired to produce a scene at once picturesque and sublime. Thomas Rowlandson, Paul Sandby Munn, John Sell Cotman, Joseph Farington and J M W Turner are perhaps the best known of the numerous artists who visited the area during this period. But one work above all others epitomizes the artist's vision of the new industrial Britain. It has become the archetypal descriptive of the first industrial nation.

Coalbrookdale by Night, painted in 1801, is the work of Philippe Jacques de Loutherbourg (1740–1812). Born in Strasbourg, the son of a painter of miniatures who moved to Paris when he was still a boy, he settled in England in 1771 and was to have a profound influence on the new generation of artists at the end of the eighteenth century. His early reputation as a designer and painter of stage sets at the Drury Lane Theatre was further enhanced in 1781 when he opened the Eidophusikon in Lisle Street, Leicester Square, a precursor of the popular Panorama and a very real ancestor of the modern cinema. Dramatic effects were produced by manipulating coloured glass in front of the oil lamps that illuminated the scenery while clouds painted on translucent linen moved diagonally past.

The powerful visual drama of the new industries had a special fascination for de Loutherbourg. He was drawn inevitably to Coalbrookdale. In *Coalbrookdale by Night* the buildings of an ironworks are silhouetted against the glare from the pig bed as the furnace is tapped. In the foreground horses pull a wagon on a plateway amongst great castings. The precise setting for the picture was for long uncertain but comparisons with contemporary views, notably a watercolour of 1803 by Paul Sandby Munn, demonstrate beyond all doubt that the scene is of the Bedlam Furnaces. Although Bedlam Furnaces are downstream of the Iron Bridge, at the time of de Loutherbourg's visit the whole area of what is now called the Ironbridge Gorge was still known generically as Coalbrookdale. Today the name is restricted to the small side valley of the Severn where Darby established his ironmaking enterprise and where the Coalbrookdale Company still makes iron castings.

NC

PORTSMOUTH BLOCK-MAKING MACHINERY

In the early history of mass production, this machinery for making ships' pulley blocks, designed by Marc Isambard Brunel (1769–1849) and built by Henry Maudslay (1771–1831), takes pride of place.

Earlier attempts at the mechanization of production were mostly limited to the making of rather simple goods, such as 'cards' for preparing cotton fibres for spinning. The usual approach to the production of an article in quantity was to subdivide the work into a large number of small manual operations which were performed at amazing speed by workers each of whom specialized in one branch.

In 1800 (when Britain was at war with France), the Navy's requirement for ships' blocks was very large, about 100,000 a year. It is not surprising that the Navy's larger suppliers had already introduced a considerable degree of mechanization. Blocks are simple enough in themselves but their manufacture involves over twenty separate operations. What was new about Brunel's proposal was the concept of a series of machines performing those operations with no need for skilled labour, leaving to human hands the moving of pieces and placing them in the machines, the working of levers, and the final assembly. Brunel devised special-purpose machines, the number of each type chosen to ensure a steady flow of components through the several operations from raw materials to assembly.

Brunel had been a French naval officer, whose royalist sympathies caused him to flee to America following the French Revolution. There he practised as an engineer, becoming Chief Engineer of New York. He came to England in 1799 to marry Miss Sophia Kingdom, whom he had met in France. While in America a chance conversation had turned his mind to the problem of the manufacture of blocks by machinery, and by 1801 his ideas were far enough advanced to take out a patent.

Through his brother-in-law, who was Under Secretary to the Navy Board, Brunel first offered the use of his invention to Messrs Taylor of Southampton, the Navy's principal supplier of blocks. Samuel Taylor's response was at once complacent and harassed, suggesting that the firm's capacity was so stretched that they could not afford time to consider improvements. Brunel was then introduced to Sir Samuel Bentham (1757–1831), Inspector General of Naval Works, who had himself studied the application of machinery to woodworking. Appreciating the merit of Brunel's scheme, Bentham persuaded the Navy to set up its own manufactory at Portsmouth Dockyard.

Brunel's machinery called for superior workmanship, and in the first instance he required working

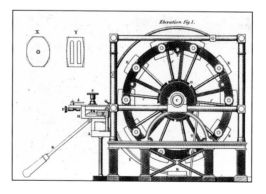

The shaping engine:
an engraving from Rees' *Cyclopaedia*, 1819

models. Another French emigré, M. de Bacquancourt, had already recognized Maudslay's exceptional ability and made the vital recommendation (*See* MAUDSLAY'S SCREW-CUTTING LATHE). Brunel first showed Maudslay a drawing of one machine without disclosing its purpose. Later he showed a drawing of another machine, at which Maudslay exclaimed, 'Ah! now I see what you are thinking of; you want machinery for making blocks'. It should be mentioned that the ability to read a drawing at all was not then common among working men. So Brunel found the high-class mechanic that he needed, and Maudslay got the contract that launched his career.

Whether the final designs were wholly Brunel's or included a significant input from Maudslay has been the subject of argument. Certainly the block machinery was highly advanced in that it comprised large machines built entirely of metal. This gave the rigidity which allowed Maudslay to build in great precision of action. As a consequence the machines ran well for a century and a half. Previously, all but the smallest machines had wooden framework, and Brunel's much publicized tools set a new standard for engineers to emulate. At the time, few other engineers could have made this step.

Brunel's new machines incorporated many radical innovations. For example the morticing machine worked automatically, and was the forerunner of both woodworking machines and slotting machines for metal cutting. The shaping engine now in the Science Museum rounded the outsides of ten blocks with a cutter moved by hand but held firmly in the machine and guided by templates.   MTW

TREVITHICK'S HIGH-PRESSURE ENGINE

The steam engines of the eighteenth century, the engines of Newcomen and of Watt, were low-pressure machines. This meant that their power output was small in relation to their size. A typical small Watt rotative engine (*See* BOULTON AND WATT ROTATIVE ENGINE) had an output of about ten horsepower (roughly that of a small motor car), and with its boiler was as big as a substantial family house. Moreover it depended on a continuous supply of cold water for condensation.

In all these engines, steam fills the cylinder and is then condensed by cooling to cause a 'partial vacuum', a drop in pressure, which gives rise to the working force. The idea of using a much higher steam pressure to push the piston without bothering with condensation was not new: it was described by Jacob Leupold in his *Theatrum Machinarum Hydraulicarum* (1725) and again by Watt in his patent specification of 1769.

But to contain and use high pressure called for an advance in engineering skill and an engineer of bold determination to see it through. Richard Trevithick (1771–1833) was just such a man. Brought up among the Cornish mines, he followed his father's occupation as an engineer. A big man who delighted in feats such as hurling the smith's sledge-hammer over the engine house, what he lacked in formal schooling he made up for with an intuitive grasp of mechanics.

In their patent specification of 1802, Trevithick and Andrew Vivian outlined both the simple high-pressure engine and some possible applications. Being much smaller than the earlier engines and not needing to be connected to a supply of cold water, it was obviously more applicable to driving road carriages and railway locomotives; Trevithick immediately began experimenting with both types of machine.

The most typical application was, however, for stationary tasks such as pumping, winding or driving machinery. Being self-contained and requiring only the simplest of foundations, Trevithick's engine was quick, convenient and cheap to install, even for temporary work such as excavation for building works. For this reason, despite its great weight, it was often known as a 'portable' engine. If users noticed that it did not live up to the exaggerated claims for economy of fuel in the advertising material, that did not prevent its widespread use; nor, in the long run, did the disastrous explosion of an early machine (through gross mismanagement) at Woolwich.

The engine in the Science Museum is a small example, of three horsepower. Even so, it is easy to see how much more compact it is than earlier types. The boiler shell, of cast iron about 40 mm thick, forms a framework to which the other parts are

Fig.1.

A Trevithick high-pressure engine used in a dredger: engraving from Rees' *Cyclopaedia*, 1819

fixed. The cylinder is fitted vertically into the boiler, an arrangement which keeps it hot and reduces the steam passages to a minimum. Steam is distributed to either end of the cylinder, where it pushes the piston up and down, and from there to the exhaust by a simple plug cock. The exhaust steam passes through a feed-water heater before being released into the chimney. The fireplace is in the mouth of a tubular flue which runs forward and then back to the chimney.

Engines like this were shipped all over the world; Trevithick himself went to set up machinery in silver mines in Peru. So perhaps there are other examples to be discovered in out-of-the-way places. But, apart from a much less complete one (also in the Science Museum collection), this is the only preserved example. By great good luck it was seen and recognized in a scrap-metal merchant's yard at Hereford in 1882 by F W Webb (1836–1906), Locomotive Superintendent of the London & North-Western Railway Company. Its destruction had already begun but Webb gathered up all the parts that could be found and had the engine restored at the company's works at Crewe. Unfortunately we have no clear record of just which parts Webb designed and made; also his restoration was nearer in time to the engine's original building than it is to our own and matches the style of the old parts well. Consequently we cannot know whether certain details which now puzzle us are the result of mistakes on Webb's part; for example, there is no obvious means of harnessing the engine to other machinery. There would probably have been a gear wheel as shown in the inset illustration, perhaps instead of the plain crank on this engine.

We can never hope to know the engine's earlier history. What we do know is that of the many firms that he licensed to build his engines, Trevithick favoured Hazledine and Co of Bridgnorth in Shropshire for the high quality of their work.   MTW

THE 'COMET' STEAM ENGINE

The earliest steam engines were too heavy to be installed in ships, but following the improvements wrought by James Watt and Richard Trevithick (*See* BOULTON AND WATT ROTATIVE ENGINE and TREVITHICK'S HIGH-PRESSURE ENGINE), steam engines were employed in both Continental Europe and North America to propel ferries and to tow barges on lakes and inland waterways. These early services were highly experimental, and accidents involving boilers caused many setbacks. Scotland's Forth and Clyde Canal was the scene for trials of a paddle tug *Charlotte Dundas* in 1801, but damage from her wash is alleged to have annoyed other canal users and she was withdrawn. Such experiments were important in demonstrating that ships need not always be dependent on favourable wind or tide.

In 1810 passenger services on the River Clyde below Glasgow were principally provided by 'flyboats', fast rowing boats whose schedules were necessarily governed by the strong tides on the river. Henry Bell was a hotelier in Helensburgh near the mouth of the Clyde and envisaged a steamboat that could bring visitors to his hotel from Glasgow whether or not the tide was favourable.

A wooden ship some forty-three feet long was built to Bell's order at John Wood's shipyard at Port Glasgow in 1811. Based on a typical sailing packet of the day, she was named *Comet* to commemorate a comet seen in Scotland that year. Steam engine design at that time was directed principally towards stationary applications and the engine which Bell ordered for his vessel might equally have powered a small workshop. When installed in the *Comet's* wooden hull, the engine's weight was balanced by that of a brick-mounted boiler on the starboard side, with a tall funnel between them doubling as a mast. Launched on 24 July 1812, Bell's twenty-five-ton steamer commenced a service between Glasgow and Greenock on Tuesdays, Thursdays and Saturdays, returning on the intervening days 'to suit the tide'; evidently she was insufficiently powerful to oppose the tide. Replacement of her original eleven-and-a-half-inch-diameter steam cylinder with one an inch larger helped to uprate the power.

Despite opposition from the flyboat owners and ridicule from some sceptics, *Comet* continued her service on the Clyde, and within two months of her launch several rival steamers were being built. Hoping that increased capacity would help the *Comet* pay her way, Henry Bell lengthened her hull by twenty feet on the beach at Helensburgh, retaining the engine provided by John Robertson. After a period of service on the Firth of Forth, the *Comet* established a passenger service between Glasgow and Fort William. Although the Hebridean islands provided shelter over part of the route, Bell must soon have realized that the *Comet* was grossly underpowered for such a service. He made up his mind to order a more powerful vessel, and on 15 December 1820 was returning aboard the *Comet* from Fort William to raise capital for it in Glasgow when a combination of wind

John Robertson with the *Comet's* engine; its original cylinder is in the left foreground

and tide forced the *Comet* ashore on Craignish Point. Bell and other passengers scrambled ashore but the *Comet*, weakened by lengthening, broke in half.

Henry Bell still owed money to various people in Glasgow and Helensburgh. One creditor more determined than most was Bailie McLellan, a Glasgow coachbuilder and founder of the city's great art collection. He took a cart up to Craignish Point, recovered the *Comet's* engine from the rocks where it had lodged and put it to work driving machinery in his Glasgow coachworks. In 1836 the engine was moved to another Glasgow factory, where its historical significance as Europe's first successful marine steam engine was recognized. It was exhibited at the Glasgow Polytechnic, but when that building caught fire in 1855, the *Comet's* engine fell from the top storey and was buried among the debris.

Six years later the engine belonged to a firm which had recently gone bankrupt, but the liquidators were unwilling to part with it. Determined to secure the engine for the nation, Bennet Woodcroft, Superintendent of Patents, turned to an old acquaintance, Thomas Lloyd, by then Engineer in Chief of the Royal Navy. He decided which Clyde shipyards should receive orders for new vessels. Woodcroft's memorandum to his Curator on the role Lloyd might play in securing the engine is a masterpiece of thinly veiled coercion:

If the firm who belong to the Comet are engine-makers, you could do no better than induce Mr Lloyd of the Admiralty to write for the engine in addition to Mr Russell. Those who make engines for the Admiralty wish to continue to do so. Those who do not would be very glad to begin.

Rising to this challenge, the Glasgow engine-builder John Napier eventually succeeded in getting custody of the engine, and it was cleaned and reassembled with a respect for authenticity rare at that time. They also took the trouble to track down John Robertson, by then over eighty years old, and he posed for photographs of the engine in their works.　　JCR

'PUFFING BILLY'

In both name and appearance *Puffing Billy* is a classic example of an ancient mechanical relic. It is also one of the most important artefacts of the earliest days of railways as we know them today – the first steam locomotive to operate successfully on a commercial basis by the adhesion of a smooth wheel on a smooth rail. Together with sister locomotive *Wylam Dilly*, preserved in the Royal Museum of Scotland, Chambers Street, Edinburgh, these are the oldest surviving steam locomotives.

To appreciate the significance of *Puffing Billy* in railway history it is necessary to consider the development of early steam locomotives. Richard Trevithick, a Cornish engineer, was first to adapt the stationary steam engine into a machine capable of propelling itself along a self-guiding track while hauling wagons (*See* TREVITHICK'S HIGH-PRESSURE ENGINE). His first design, built at Coalbrookdale in 1802, was inconclusive but his second in 1804 demonstrated at Pen-y-darren in South Wales the potential of a steam locomotive to replace horses for the haulage of coal wagons. The boiler on these early designs was comparatively large and heavy. While this weight helped adhesion it led to many broken rails.

News of the success of these limited trials and their significance for colliery working in particular soon spread. Christopher Blackett, the owner of Wylam Colliery in Northumberland, was interested in the potential of the Trevithick designs to replace horses on the five-mile-long wooden wagonway between his colliery and the head of navigation on the River Tyne. Blackett was deterred, however, from concern that the weight of a locomotive might damage the wooden rails.

Early in 1812 Blackett's interest was revived after the successful operation of a more advanced design of locomotive, with two vertical cylinders, on a rack railway at Middleton Colliery near Leeds. Here a toothed wheel on the locomotive engaged similar teeth on a continuous rail, thus enabling steep gradients to be overcome. There were, however, difficulties and limitations at junctions in the track.

The quickening pace of industrial growth placed a premium on productivity and this, together with the increasing price of horse fodder during the Napoleonic Wars, further encouraged the replacement of the horse by more efficient motive power. The relatively level Wylam Wagonway had however been relaid in 1808 with cast-iron plateway, to the same five-foot gauge as the previous wooden rails. Rather than incur the cost of relaying again with the more expensive and limited rack rails Blackett decided to build a steam locomotive that could operate on smooth iron rails. The challenge was to design one that would provide sufficient adhesion to pull a commercial load on a conventional wagonway without an unacceptable rail breakage rate.

Accordingly in October 1812 Blackett asked his colliery manager William Hedley to build such a locomotive. This was constructed during 1813, probably in Gateshead, with a single cylinder and a flywheel. It failed in trials primarily through lack of steam, having only a single heating flue in the boiler. A further locomotive, with two cylinders and a return flue was built, probably at Wylam, sometime during the winter of 1813–14. This locomotive apparently entered service in March 1814 and could successfully haul up to nine or ten coal wagons at speeds of 4–5 mph on the Wylam Wagonway. This is the locomotive which became known as *Puffing Billy*.

It was followed later in 1814 by two more locomotives subsequently known as *Wylam Dilly* and *Princess Mary*, although these names were never actually carried by the locomotives themselves. Contemporary reports mention the strange noise made by the engines and, indeed, there is evidence on *Puffing Billy* that the exhaust steam passed through a chamber on top of the boiler designed to reduce this noise. The name *Puffing Billy* is a relatively obvious nickname, 'Billy' being a corruption of the word 'dilly' which was a local name for the caldron or coal wagon used on the Wylam Wagonway.

While the locomotive as built was successful, there were obviously still problems with damage to the plateway. Contemporary sketches show the original four-wheel locomotive modified to an eight-wheeler to spread the weight. At some later stage, perhaps when the plateway was rebuilt in about 1828, with stronger cast-iron edge rails but still to the five-foot gauge, the locomotive reverted to its original four-wheel configuration which it then retained. The flanged wheels it now possesses would no doubt have been added at this time. In other respects the locomotive appears very much as built, although comparison with early drawings indicates that at some stage the valve gear had been modified.

Puffing Billy continued working at Wylam until the early 1860s. The modest task for which the locomotive was intended meant that its performance was adequate for many years. This is in contrast to many of its illustrious successors such as *Rocket*, (*See* STEPHENSON'S ROCKET), where the pace of development of locomotives at the forefront of railway evolution led to a much shorter working life, at least at the work for which they were originally built.

Despite the success of *Puffing Billy*, much remained to be done before the full potential of the railway could be realized. Perhaps the greatest underlying significance of *Puffing Billy* is that its success encouraged George Stephenson in his interest in steam locomotives and railways. It was he who was to develop the steam locomotive to the point at which the success of the railways, as we know them today, was proven.

JAC

DAVY'S SAFETY LAMP

These lamps, designed by Sir Humphry Davy (1778–1829) are said to be the first safety lamps actually used in a coal mine.

In the eighteenth century miners worked by the light of candles or other naked flames. As mines became deeper a new hazard appeared – firedamp. This is a gas, composed mainly of methane, produced by the decomposition of vegetable matter when coal seams are formed and retained in fissures and molecular interstices. The gas is not poisonous but when mixed with air in certain proportions it is explosive and can be ignited by a naked flame.

An increasing number of miners lost their lives in these pit explosions and several attempts were made to develop a lamp that would be safe underground. Carlisle Spedding (c 1696–1755) invented a steel and flint mill where light was obtained in the form of a stream of sparks created by rotating a thin steel disc against flint. This was quite widely adopted in the north of England and was safer than a candle. However, explosions still occurred and a boy was needed to operate it so it was expensive.

On 28 May 1812 an explosion at Felling Colliery killed ninety-two men and boys. The publicity given to it by the vicar of the parish, the Reverend John Hodgson, resulted in the establishment, on 1 October 1813, of the Sunderland Society for Preventing Accidents in Coal Mines. The Society approached Sir Humphry Davy for help. Davy met John Hodgson and John Buddle, the leading mining engineer of the day, and visited Hebburn Colliery where he carried out some experiments. After Davy returned to London Hodgson sent him some firedamp in six wine bottles and shortly afterwards Davy wrote,

I have already discovered that explosive mixtures of mine damp will not pass through small apertures or tubes; and that if a lamp, or lanthorn be made air tight on the sides, and furnished with apertures to admit the air, it will not communicate flame to the outward atmosphere.

Sir Humphry Davy

Subsequently on 9 November Davy read a paper to the Royal Society entitled 'On the firedamp of coal mines and on methods of lighting the mines so as to prevent its explosion'. Three lamps were described in the paper. They were glass lanterns but with the air feed through concentric metal cylinders or, in one case, through wire gauze. The top of the glass chimneys was similarly equipped. By this time he had discovered the principle on which the safety lamp was to be constructed but it was another month or so before he conceived the idea of surrounding the flame of the lamp with wire gauze.

The first Davy safety lamps were complete at the end of 1815 or early 1816 and sent to John Buddle. They were tested in Hebburn Colliery on 9 and 17 January. John Buddle recorded his reactions:

To my astonishment and delight, it is impossible for me to express my feelings at the time when I first suspended the lamp in the mine, and saw it red hot; if it had been a monster destroyed I could not have felt more exultation than I did. I said to those around me, 'We have at last subdued this monster.'

Each lamp has a cylinder of brass wire gauze of one and a half inches in diameter and five inches high screwed to the top of a reservoir of oil. In one there is also a spout for filling the vessel and a pricker to trim the wick.

Davy was well aware that in certain circumstances, such as a strong current of firedamp, his lamp could be unsafe and he himself suggested some modifications. These included a gauze cap over the top of the main gauze to cool the burnt gases and a tin shield sliding on the frame wires of the lamp to protect it from draughts.

The Davy lamp was widely adopted and over the years many modifications and improvements were made. Davy was awarded a service of silver plate valued at £2,500 in recompense for his efforts. Unfortunately, the Davy lamp did not reduce the number of accidents and fatalities: it merely allowed deeper and less safe mines to be exploited.

At about the same time as Davy was engaged in his experiments, George Stephenson (1781–1848) was working on a safety lamp also based on the principle of explosive gases not passing through small tubes and apertures. There arose considerable controversy, never fully resolved, about whose was the first safety lamp. In practice Davy's lamp was far more significant because it was his understanding of the scientific principles and application of the wire gauze that was so valuable and which formed the basis for the design of virtually all flame safety lamps. Indeed Stephenson also soon adopted a wire gauze for his lamps.

In 1830, John Hodgson gave these early Davy lamps to a friend by whom they were presented to the Museum of Practical Geology. They were transferred to the Science Museum in 1895.   JK

LISTER'S 1826 MICROSCOPE

It has been suggested that this is possibly the most important optical microscope ever made. It was designed by Joseph Jackson Lister (1786–1869), father of the famous surgeon Joseph Lister, in March 1826, and made by William Tulley, the London optical instrument maker. It included such improvements as graduated draw tubes (to make it easier to set up and focus), lenses to act as a sub-stage condenser (increasing the amount of light going through the microscope slide), and a rotating and clamping stage to manipulate the slide. Its main significance rests on the superb quality of the objective lens (the one nearest the specimen), from which distortions and colour change effects had been eliminated to a greater degree than ever before.

Joseph Jackson Lister's interest in optical matters can be traced back to his schooldays, and during the rest of his life he designed and tested lenses and other microscopical equipment. Despite being a superior student in schools at Hitchin, Rochester and finally Compton in Somerset, Joseph Jackson left formal education at the age of fourteen to join his father's wine business. His own talents and his father's generosity were demonstrated four years later, when in 1804 he became a fully-fledged partner. In 1821 he became part owner of a ship commanded by his brother-in-law, and this and other business interests continued throughout his life underpinning his scientific and religious activities.

He was an active and influential Quaker and became a school visitor and trustee. It was through his educational activities that he met his wife Isabella, who taught reading and writing until the day of her marriage.

Joseph Jackson was elected to fellowship of the Royal Society in 1832, on the strength of his work in microscopy and histology (the study of tissues). Of his seven children, only his son Joseph was to reach or surpass his position; the oldest son John died of a brain tumour in 1846 at the age of twenty-four. In those days it was quite unusual for all the children in a family to reach adulthood, and Joseph Jackson never lost sight of the wider world of suffering. In 1848 he not only served on a committee to raise funds for the relief of Irish potato famine victims but took the trouble to visit them to discover for himself the conditions under which they were living.

As a result of Lister's developments in lens design, the true appearance of tissues could be accurately described for the first time. Blood corpuscles were seen to be concave discs rather than globules, and muscle was shown to be made of fibres, with cross striations, irrespective of the source of the muscle. In the spring of 1827, Dr Thomas Hodgkin published jointly with Lister the results of these and many other observations on nerves, arteries, cellular membranes and the brain, refuting the views of some of the most eminent researchers of the time.

Lister published his further work on objective lenses in 1830, in the *Philosophical Transactions of the Royal Society*, and established the compound microscope (one with two main lenses) as a serious research instrument for the first time. He also continued to use it himself, and from the end of 1830 was making his own lenses, despite having no previous experience. His own research interests extended into investigations of the limits to human vision, and another paper on zoophytes (microscopic animals) illustrated by his own drawings from the camera lucida, an attachment which allows the microscopist to view both the specimen through the microscope and a sketch pad alongside at the same time, so outlines can be 'traced' directly on to the paper.

He continued to take an interest in the design of the microscope, advising the London makers Andrew Ross, and later, James Smith, on elements of lens design. With both these men the association was long and commercially invaluable, some of Lister's improvements being widely adopted as standard.

Because of his naturally diffident nature, much of Lister's work was never published under his own name. The full width of his many interests has only recently begun to come to light with a detailed study of his drawings, many of which were given to the Royal Microscopical Society on the death of his son Joseph, and the true significance of his work can now be more widely appreciated.   JEI

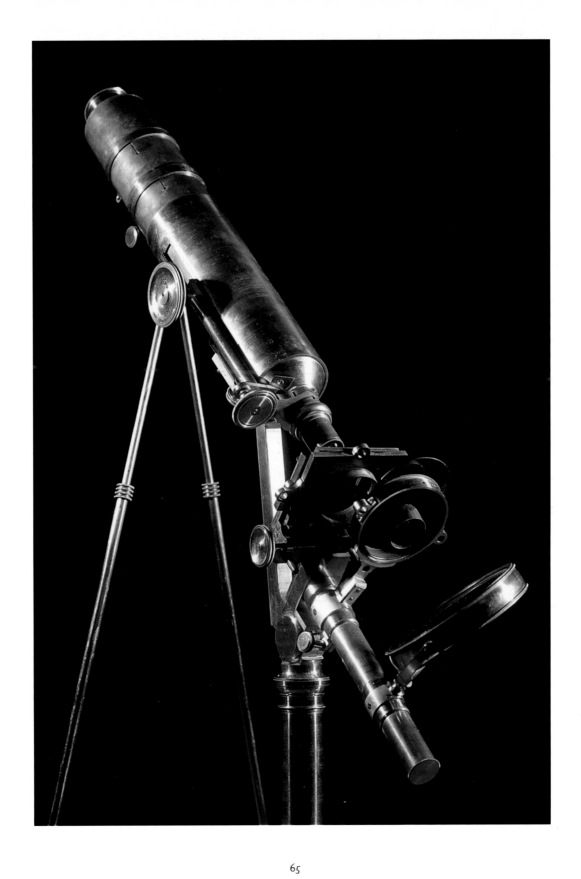

BELL'S REAPER

The machine which has come to be seen as the first truly practical mechanical reaper was designed in 1827 by Patrick Bell (1799–1869), later Reverend Bell, of Carmyllie, Scotland. It incorporated several features that are still seen in the modern binder, stripper and combine harvester. Yet Bell's pattern was never produced in large numbers, and it was to be the success of American designers, most notably Cyrus McCormick, at the Great Exhibition of 1851, that would lead eventually to the establishment of mechanized reaping in Britain. Only then was the Reverend Bell given the opportunity to assert his importance in the history of the reaper's invention.

Reaping (the cutting of the corn at harvest) had for centuries been achieved using hand-held sickles and scythes. However, a mechanical reaper is described in the writings of ancient Roman scholars and in 1787 a depiction of a machine based on these accounts appeared in the periodical *Annals of Agriculture*. The designer made no attempt to build it, but by the end of the eighteenth century several people had constructed reapers based on the Roman model. All of these proved unsuccessful.

Some advances were made in the early years of the nineteenth century, for example the full-size machines by such pioneers as Thomas Plucknett, Joseph Mann and James Smith of Deanston (1789–1850). Most were ingenious, but were eventually abandoned. It was Bell's reaper that proved to be the first workable design.

Patrick Bell was the son of a farmer. He showed an early interest in mechanics and for some years pondered ways to mechanize reaping in order, he later explained, to ease the burden of his father's workers at harvest time. Ignorant of previous research, except that by James Smith (above), he dismissed a succession of ideas until he began work on a design incorporating the cutting action of garden shears. A small working model was made in 1827 followed by a full-scale prototype.

Bell was very secretive about his invention, and went to extreme lengths to ensure the early trials of 1828 were unobserved.

Towards the end of 1828 it was successfully exhibited at a small number of venues. A specially commissioned reaper was also shown before the Highland Society where Bell was awarded fifty pounds (immediately paid to a creditor). Further exhibitions in 1829 were sufficiently successful for a few machines to be made for interested parties.

Wider interest was slow to develop and by the mid-1830s less than twenty machines were in use, mainly in Scotland. As the reapers required regular maintenance, many gradually fell into disrepair when in the hands of inexperienced users. A further factor that led to their decline was Bell's reluctance to patent his idea. This resulted in varying standards of workmanship on his own reapers, and also allowed inferior copies to be made by other manufacturers.

International sales were negligible, but it has been suggested that several of his reapers were exported to North America in the early 1830s; certainly drawings were published there. The influence this had on the American designers who were to dominate the market proved a controversial issue.

Widespread recognition was belatedly afforded to Bell after the Great Exhibition, where the American machines of Cyrus McCormick and the Hussey Company had created a stir. Now powered by two horses and with an improved cutting mechanism, his reaper emerged from obscurity to compete against the American models, in a series of trials that heralded the acceptance of the reaper in this country.

Thus the real relevance of Bell's reaper was finally recognized, although it was not until 1867, only two years before his death, that he received a testimonial of £1,000 from the Highland and Agricultural Society of Scotland, in appreciation of his contribution. The reaper was acquired by the Patent Museum from Bell himself in 1868, in exchange for a more up-to-date model.   SRE

Bell's Reaper: engraving from the *Rural Cyclopaedia*, 1847

STEPHENSON'S 'ROCKET'

In the history of the early development of the steam railway locomotive there is no more significant machine than *Rocket*. Its design and construction produced a momentous leap forward in locomotive and railway practice as it embodied the fundamental principles which were followed through the 150-year history of the reciprocating steam locomotive. It clearly demonstrated the superiority of moving locomotive engines over stationary engines and rope haulage systems for railway transport. Yet *Rocket* was an experimental machine at the forefront of a rapidly developing technology. As a result it was soon modified and rendered obsolete in little over a decade.

The first steam locomotives were built for hauling heavy loads rather than for speed (*See* 'PUFFING BILLY'). They offered little advantage over stationary engines and rope haulage. The directors of the Liverpool & Manchester Railway (L&MR) required a faster and better form of motive power for their inter-city railway. They offered a premium of £500 for a 'Locomotive Engine which shall be a decided improvement on those now in use'. Competing engines were to be tested on a length of the new railway at Rainhill, east of Liverpool.

Rocket was built at the Robert Stephenson and Company locomotive works in Newcastle-upon-Tyne to compete for the premium at the Rainhill Trials of October 1829. It was the eighteenth steam locomotive built by the company which combined the experience of George Stephenson (1781–1848) and his son Robert (1803–59). It was Robert who undertook the construction of this innovative machine: a lightweight locomotive with a single pair of driving wheels and built for speed, as its name implied.

In the trials *Rocket* was in competition with four other machines and emerged the clear winner. The locomotive's success was largely due to the incorporation of a number of important design features. The multi-tube fire tube boiler had been suggested to the Stephensons by Henry Booth and was greatly superior to the more usual single or twin flue boiler

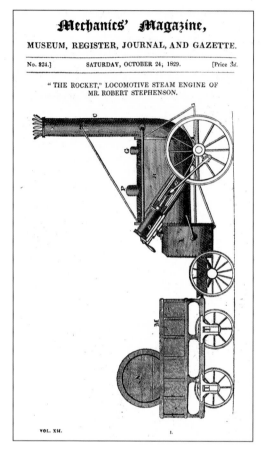

Mechanics' Magazine,

MUSEUM, REGISTER, JOURNAL, AND GAZETTE.

No. 824.]   SATURDAY, OCTOBER 24, 1829.   [Price 3*d*.

" THE ROCKET," LOCOMOTIVE STEAM ENGINE OF
MR. ROBERT STEPHENSON.

VOL. XII.   I.

The *Rocket* in its original form:
frontispiece of the *Mechanics Magazine*, 24 October 1829

barrel. Twenty-five three-inch-diameter tubes in the boiler increased the heating surface and as a result the amount of steam the boiler could produce. This was further assisted by use of a water jacket around the firebox. Steam from the cylinders was exhausted through a blast-pipe up the chimney which gave induced draught in the firebox and made the fire burn more fiercely. *Rocket* was also built with steeply angled cylinders which drove directly to a single pair of driving wheels with crank pins at right-angles; a more effective system than the vertical cylinders of

earlier engines. The multi-tube boiler, blast-pipe exhaust and the two-cylinder simple drive mechanism are the three fundamentals of steam locomotive design which were first combined in *Rocket*.

Although built expressly to compete for the first prize and not for everyday traffic work, *Rocket* so impressed the directors of the L&MR that they purchased it for use on the line and ordered four similar locomotives from Robert Stephenson and Co. *Rocket* assisted with works trains during construction and on 15 September 1830 took part in the opening ceremonies during which it was involved in an accident in which William Huskisson, then MP for Liverpool, was killed. This was the first recorded passenger fatality on a railway.

Within eighteen months of its construction *Rocket* had been significantly rebuilt. The cylinders were swapped around and dropped to eight degrees from the horizontal to improve the riding of the locomotive and the splayed-out chimney base was replaced by a more practical drum-shaped smokebox. Other changes included the addition of a buffer beam and a reduction in the chimney size.

By the mid-1830s 'Rocket-type' locomotives and their successors had displaced the prototype on the L&MR. *Rocket* was sold in 1836 to Messrs Thompson of Kirkhouse near Carlisle. It had a three-year working life on the Brampton Colliery Railway in Cumberland but by about 1840 it was considered too lightweight and worn out to work coal trains and was laid aside.

A proposal to show the locomotive at the Great Exhibition of 1851 came to nothing, and the derelict locomotive was presented by Messrs Thompson to the Patent Office Museum in 1862. For public display Stephenson & Co undertook some restoration work on *Rocket*. This included the addition of parts incompatible with the rebuilt condition of the locomotive. Some of these have since been removed. What remains are mainly those parts believed to date from 1829–36 and represent the bare bones of the locomotive in its rebuilt form.   DWH

BABBAGE'S CALCULATING ENGINES

Charles Babbage (1791–1871) was an English gentleman of science with an impressively wide range of interests. He was an inventor, reformer, mathematician, philosopher, scientist, critic, political economist and a prolific writer. He is nowadays widely known for his pioneering work on vast mechanical calculating engines. The few mechanical assemblies that survive represent the earliest antecedents to the modern computer and are among the most celebrated icons in the prehistory of computing.

The tedium of doing calculations by hand had stimulated attempts to build mechanical calculators well before Babbage's own efforts and the task had challenged some of the most versatile intellects of all time. Devices were built in the seventeenth century by Blaise Pascal, the French philosopher-mathematician, and by Gottfried Leibniz. Pascal's invention was paraded before the social and intellectual élite of the day and caused something of a stir. Though these early desk-top devices represent significant steps in the evolution of mechanical calculators they were more in the nature of ornate curiosities.

Before the widespread availability in the twentieth century of dependable mechanical, and, later, electronic calculators, scientists, astronomers, navigators, actuaries, bankers and the like relied for the most part on printed mathematical tables to perform calculations requiring more than a few figures of accuracy. The repetitive calculations for these tables were performed by hand by people called 'computers' and results were then copied and set in loose type for printing. Inevitably there were mistakes.

The task that seized Babbage in 1821 was to build automatic calculating machines that would eliminate all these sources of inaccuracy at a stroke. The 'unerring certainty of mechanism' would free calculation of human error, and having the machine print the results automatically would eliminate the risk of mistakes in manual transcription and typesetting. Babbage's Difference Engine was conceived to do just this – to calculate and automatically print error-free mathematical tables. The Difference Engine is so called because of the mathematical principle upon which it is based, the method of finite differences. The advantage of the method is that it allows certain complex mathematical expressions to be calculated using simple addition only, without the need for multiplication and division which would ordinarily be required.

Work on Difference Engine No.1 started in the early 1820s. The project was abandoned in 1833 after a dispute with the engineer Joseph Clement who had been engaged to build the machine. About one-seventh of the machine had been assembled as a demonstration piece by Clement in 1832. This portion of the engine is the first known automatic calculator and is one of the finest examples of precision engineering of its day. It is automatic in the sense that for the first time, mathematical rule was successfully incorporated in mechanism: the operator did not need to understand the mechanical or logical principles to achieve useful results. All that was required was to turn the handle and the machine did the rest.

The Difference Engine is capable of a fixed set of operations determined by its wheel work, ie. it is not a general purpose machine and we would nowadays prefer to call it a calculator rather than a computer. Babbage's Analytical Engine, however, conceived by 1834, has features that are startlingly similar to those of a modern electronic computer. The Analytical Engine was programmable using punched cards – a technique used in the Jacquard loom to control patterns woven with thread. The engine had a repertoire of basic operations (multiplication, division, addition and subtraction) and could automatically execute sequences of these operations in any order. The internal organization of the machine was also surprisingly modern in conception: the 'mill', where information was processed, was physically separate from the 'store' or memory where information was kept. The separation of 'store' and 'mill' (today called the central processor) is a feature that has dominated the design of the electronic computer since the mid-1940s.

In operation the engine was capable of 'looping' – repeating the same sequence of operations a specifiable number of times, and of 'conditional branching' – choosing one of a number of alternative actions depending on a predetermined condition being met. While we tend to speak of the Analytical Engine as though it was a physical thing, the machine was never actually built though the designs were highly developed. It would have been the size of a small locomotive. All that Babbage accomplished in the way of hardware was a small, simplified experimental model of a portion of the 'mill' which was under construction at the time of his death in 1871. This assembly does scant justice to the grandeur of his schemes but remains nonetheless a monument to the earliest attempts to realize an automatic general purpose computing machine. DDS

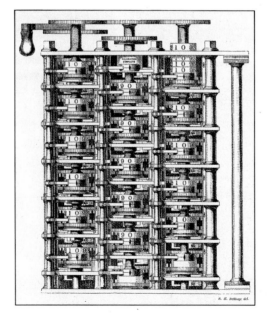

Above: Portion of Difference Engine No 1 assembled in 1832; woodcut, 1853.
Opposite: Experimental model of a portion of the Analytical Engine mill, 1871

70

TALBOT'S 'LATTICED WINDOW'

What is generally regarded as the earliest existing photographic negative measures only 28 × 36 mm. It is mounted on to a wafer thin piece of black paper, which in turn is mounted on to a sheet of writing paper, on which are written the words:

Latticed Window (with the Camera Obscura)
August 1835.
When first made, the squares of glass about 200 in number
could be counted, with help of a lens.

This label, the mount, and diminutive negative are the work of William Henry Fox Talbot (1800–77) of Lacock Abbey, Wiltshire. Henry Talbot (not, as popularly supposed, Fox Talbot) was not satisfied to follow the role model established by those English gentry who managed their estates and generally led lives of pleasure. Instead Talbot pursued ideas, using the patronage offered by his inheritance.

Talbot's 'idea' for photography came from disappointment with his ability to draw Italian scenery using the 'camera lucida'. This was a portable instrument which used a small prism to project an image of a scene on to a piece of paper. He began to consider the agencies of optics, light and chemistry in an attempt to capture the elusive view.

The essential elements had long been available. Chemists in the eighteenth century knew that the salts of silver darkened with exposure to light. The 'camera obscura' was made by opticians of the seventeenth century for use by artists interested in achieving correct perspective. (Meaning 'darkened room', this was a light-tight box in which a small hole or lens allowed light to enter so that an image of the view outside was projected on to the opposite side.) But it required a fertile imagination to bring them together and realize the potential for making a permanent record of the type of image produced by the camera obscura.

In the spring of 1834, Talbot began his first experiments at Lacock. Tradition has it that Talbot had his estate carpenter make his first cameras.

In Talbot's system, a sheet of writing paper was brushed first with a solution of common salt. A second wash of silver nitrate combined with the salt to form light-sensitive silver chloride. The prepared paper was then gummed inside the back of the camera where it was exposed until it darkened, a matter of around an hour rather than minutes. Once the image (in which light and dark areas were transposed) was visible it was removed from the camera and stabilized against the further action of light by being 'fixed'.

Fixing the images was a key element in the success of Talbot's idea. Earlier experimenters had made images using much the same chemical methods as Talbot. But all had failed to find a suitable fixative and their images disappeared once exposed to daylight. Talbot's initial answer was to fix his prints in a strong

An enlarged positive print from
Talbot's 'Latticed Window' negative

solution of common salt which rendered them insensitive to light.

Equally important was Talbot's grasp of the concept of a negative being used to make a positive. He noted in February 1835 'if the paper is transparent, the first drawing may serve as an object, to produce a second drawing, in which the lights and shadows would be reversed'. The original appearance of the image would be restored, and multiple copies produced – a description of photography that serves as well today as it did then.

Events were unkind to Talbot. In France, independently, L J Daguerre had also been struggling to make images through the agencies of light and chemistry. News of his discoveries was first reported in the *Literary Gazette* on 12 January 1839.

Talbot was placed in a quandary. He had no details of the working methods of Daguerre's process (See EARLY DAGUERREOTYPES OF ITALY). He feared it would be identical with his own and preempt any claim to priority he could reasonably make. Time was against him and speed was essential. He selected examples of work from the brilliant summer of 1835 to demonstrate 'the wide range of its applicability'. He mounted the photographs on slips of thin black paper to enhance their appearance. They were exhibited at the evening lecture of the Royal Institution on 25 January just thirteen days after the report of Daguerre's invention and were warmly received by the audience.

There are at least five surviving examples of other images similarly mounted, suggesting they also were on display at the Royal Institution that evening. However the only image to be dated and inscribed by Talbot is the 'Latticed Window'. We do not know when or why Talbot added this information. The presence of these facts has given the negative an exclusive identity although it is possible that one of the other images from the 1835 collection predates August. Constantly reproduced in histories of photography, the 'Latticed Window' has an iconic status that will survive all subsequent revisions.　RT

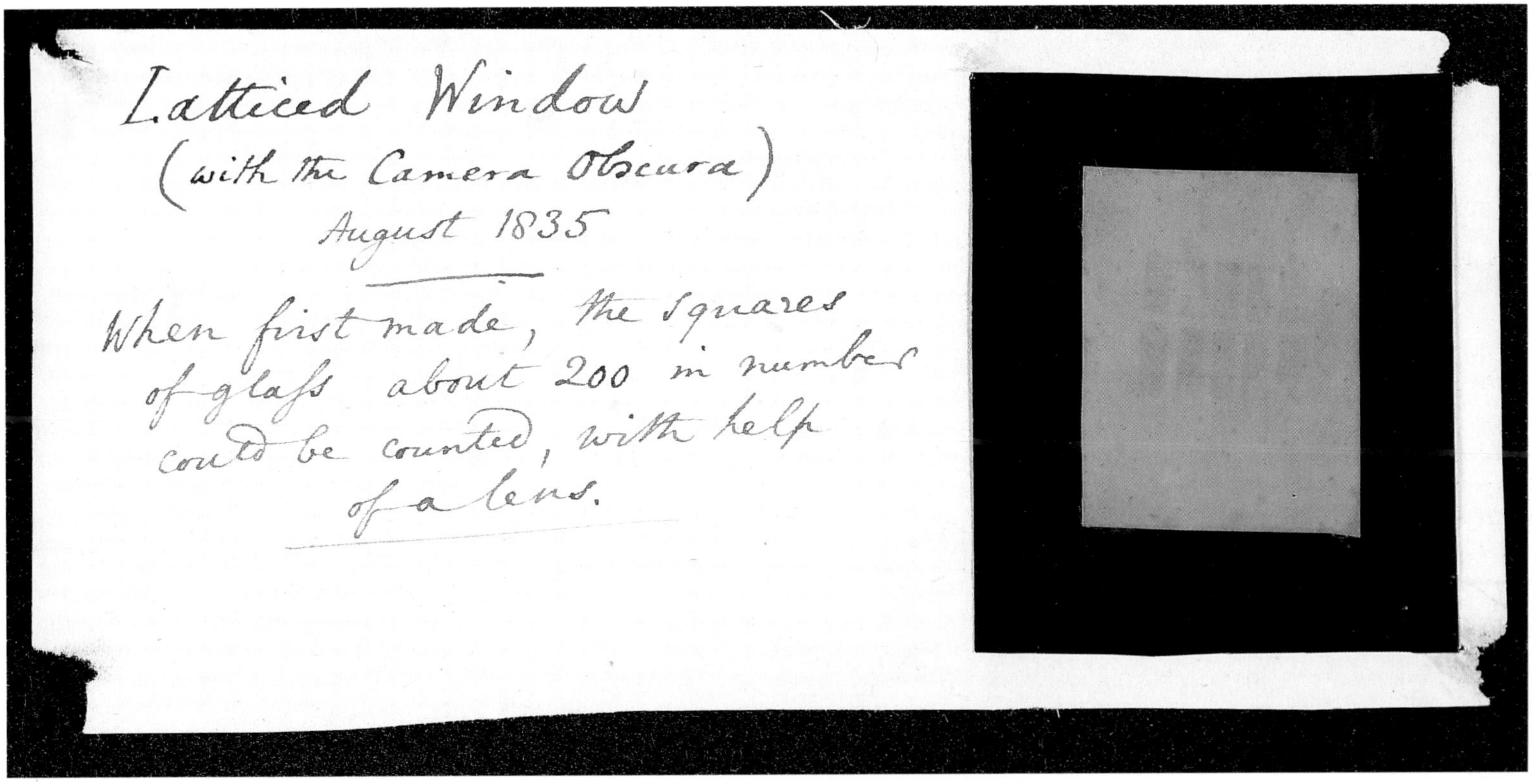

Latticed Window
(with the Camera Obscura)
August 1835

When first made, the squares
of glass about 200 in number
could be counted, with help
of a lens.

COOKE AND WHEATSTONE'S TELEGRAPH

In 1776, when the American colonies declared themselves independent of Britain, news of the event took forty-eight days to cross the Atlantic. Some sixty years later, an American called Samuel Morse found a way of eliminating such embarrassing delays. Variants of the Morse telegraph were later to link the world and, paradoxically, strengthen Britain's ties with her still expanding Empire. However the world's first public telegraph service used not Morse's invention but this instrument, patented in 1837 by Charles Wheatstone (1802–75) and William Cooke (1806–79).

Inventors had been trying since the mid-eighteenth century to use electricity for remote communication. All used various impractical means of detecting signals. Cooke and Wheatstone swept all this away. They used Oersted's recent discovery that current electricity could move a compass needle. They added the idea of a coding scheme to make five wires do the work of twenty. And they realized that the nascent railway network offered the ideal market for instant communication.

The instrument shown here used six wires for transmission. Five were used for letters of the alphabet. A letter was sent by pressing two of the twelve buttons on the transmitting instrument, causing two of the five needles on the face of the receiving instrument to deflect in opposite directions and point to one of twenty letters. The sixth wire deflected one of the needles to left or right to signal the ten digits.

The idea of using a set of compass needles whose states of deflection would form a code seems to have originated with Cooke, a retired army officer with no electrical training. However he became convinced that something less obscure was required and settled on a complicated system using a direct-reading clockwork receiver.

Cooke started hawking his invention round the new railway companies in January 1836, but by February was running into trouble getting it to work over a one-mile circuit. He had already consulted Faraday, who was helpful but cool. He was now recommended to Wheatstone, Professor of Experimental Philosophy at King's College, London, who had some experience of long electrical circuits.

By March 1837 Cooke and Wheatstone had agreed to become partners. Wheatstone clearly persuaded Cooke to return to his earlier multiple-needle idea. By June they had a patent. In July, using a four-needle prototype, they gave a demonstration to railway engineer Robert Stephenson. He ordered a pair of five-needle instruments, which by August were in service on the Euston to Camden Town section of the London and Birmingham Railway, forming the world's first operational telegraph installation. It was not a success; in October 1837, Cooke and Wheatstone offered to buy the instruments back.

Within six months Cooke had sold the invention to the Great Western Railway which was pushing out from London towards Bristol. By July 1839 the line had covered the thirteen miles to West Drayton, and a five-needle instrument was installed at each end. One of these was the instrument now in the Science Museum. The five-needle system survived until 1843, used not only by the railway but also for public telegrams – the first such service anywhere. Then, with the extension of the railway westwards, the inexorable convergence with Morse's single-wire system began.

The five-needle system offered low running costs, because it was easy to use. It could be – and was – operated by children. But the capital cost of running six wires began to tell as the line stretched out over the 120 miles to Bristol. Cooke and Wheatstone re-engineered the system to use first two needles, then one. In doing so they had to adopt a code which replaced an easily-interpreted spatial pattern by a more obscure temporal one, the sequence of left and right deflections of a singe needle. The convergence with Morse's dot-dash system was almost complete.

There is no space here for the story of Cooke and Wheatstone's opposition to Morse, or their subsequent attempt to cooperate with him in America; nor for the bitter disputes which led to the break-up of their partnership in 1845. What matters more than these conflicts and rivalries, local or transatlantic, is that Cooke and Wheatstone were the first to speed public messages beyond the pace of ship and horse. As such, they stand secure among the founders of the twentieth century.   RB

Wheatstone's inventiveness was not confined to telegraphy. He developed the stereoscope and the concertina, and also built a speech synthesizer. In the 1850s he also designed three remarkable typewriters. All reveal his background as a musical instrument maker

THE BROUGHAM

Despite the profusion of carriage styles in the nineteenth century, there were only a small number of fundamental changes to their general configuration. The most important of these was the introduction of the brougham in 1838.

The design of carriages started to diverge from that of wagons with the introduction of suspension in the mid-sixteenth century. In these early coaches there was an undercarriage in which a large wooden beam called the 'perch' joined the rear axle to the forecarriage, under which turned the front axle. At the four corners of the undercarriage the body was suspended from upright posts by leather straps called 'braces'. The suspension improved as metal springs began to replace the braces in the latter part of the eighteenth century, and on the introduction of the elliptical metal spring in 1804. But the coach body was still high-mounted above the perch.

In the brougham this was dispensed with, the body itself being all that linked the forecarriage to the rear axle. This meant that the body not only became a structural element itself but could now be mounted much lower down between the axles. Instead of a high, heavy, ponderous coach it became possible to build a low, light, manœuvrable, closed vehicle which could be drawn by a single horse. This was better suited to use in towns and was cheaper to buy and to operate. The brougham, and the clarence which developed from it, were the most popular type of closed carriage built, and influenced the design of other closed and open carriage styles during the rest of the century.

It is not known what gave Henry Brougham (1778–1868), Lord Chancellor of England, the idea for this new type of carriage. However, it is known that he collaborated with a Mr Robinson of the firm of Robinson and Cook of Mount Street, London, and that it was delivered by the firm on 15 May 1838. In fact it was not owned by Lord Brougham but 'jobbed' or contract-hired to him. Indeed he is said not to have liked it very much and to have used it only occasionally. Certainly he later had an improved one made for him by the same firm. The original brougham was sold to Sir William Foulis of Ingleby Manor, Stokesley, Yorkshire on 27 August 1840. Compared with later broughams it was rather boxy and inelegant; yet it can be regarded as a prototype, of necessity built before the proportions could be refined.

The brougham subsequently became the property of Lord Henry Bentinck who in turn sold it to Henry Earl Bathurst. It is reputed to have carried both Disraeli and Gladstone. Recognizing its historical importance, Earl Bathurst presented it to the Worshipful Company of Coachmakers and Coach Harness Makers in 1894, from whom it has been on loan to the Science Museum since 1895. Mr George N Hooper, of the Hooper coachbuilding company and a Past-Master of the Coachmakers Company, inspected the brougham with Earl Bathurst at Cirencester House on 5 April 1894. He was able to confirm its provenance by inspecting the books of the original builders and by speaking to Michael Elton, Earl Bathurst's former coachman who had known the carriage in the Bathurst establishment for over forty years. Hooper recorded his impressions of the brougham:

It differs from those now made in many points of design – proportion, construction, finish. Those now made [1894] are much more refined in outline, and more pleasing to the eye; the body is several inches wider in front than at back, and although larger and heavier, does not provide the same comfort and convenience that those of smaller and lighter construction now afford. PRM

Above: A mid-Victorian design for a brougham
by Barker & Co, London.
Opposite: The first brougham was restored in 1977
for the tercentenary of the Worshipful Company of
Coachmakers and Coach Harness Makers

EARLY DAGUERREOTYPES OF ITALY

During his visit to Venice Dr Alexander John Ellis (1814–90) rose early. Even as a novice photographer, brief experience had already taught him early morning light gave the best results. Before eight o'clock, he was busy setting up his camera, a cumbersome thing that had to be placed on a tripod. After each exposure he carefully recorded exposure times and weather conditions in his pocket notebook for future reference. There was still much to learn.

By mid-afternoon when he reached the Rialto Bridge he had already made four exposures that he felt were 'very good'. The bridge was a natural subject for photography with its historic mercantile associations. Having gained access to an upper window of the White Lion Inn, he carefully composed the picture and made ready for the exposure. Exposure times were measured in minutes rather than fractions of a second and successful results were regarded as substantial achievements. Assessing the light, he judged that thirteen minutes would give a good result. This was 20 July 1841 and the art of photography was in its infancy. Little more than two years had elapsed since Daguerre and Henry Talbot had announced their exciting discoveries (See TALBOT'S 'LATTICED WINDOW').

Ellis was a fortunate young man. Born Alexander Sharpe, he changed his name in 1825 to satisfy the terms of a bequest from a relative who wished him to devote his life to study and research. Educated at Shrewsbury and Eton, he became a scholar at Trinity College, Cambridge in 1835 where he fully immersed himself in the academic and social life of the university. After graduating in 1837, he entered the Middle Temple but appeared to have no intention of devoting his life to law. Travel was infinitely more appealing to a young man of wealth and leisure.

March 1839 found him in Dresden where news of an exciting discovery by Monsieur Daguerre of Paris was appearing in the newspapers. Sceptics could not believe that it was possible to make images by the agency of the sun alone.

Daguerre's photographic process was based upon solid metal. A highly polished sheet of silver-plated copper was made sensitive to light by exposing it to the fumes of iodine. After exposure the image was 'brought out' by further exposure to the fumes of warmed mercury. The resultant plate, known as a daguerreotype, drew gasps of amazement from an admiring audience, wherever displayed. What captured the imagination was the exquisite detail, the infinite accuracy of information, beyond anything the hand of man alone could produce.

The traditions of the eighteenth-century Grand Tour still lingered in early Victorian Britain, despite its complete abandonment during the Napoleonic Wars when the whole of Europe was effectively closed to leisured travellers. Interest in the sites of the Ancient World, seen as the cradle of European civilization, still ran strong amongst the educated classes. It was a necessary part of their upbringing to understand the history and culture of these early civilizations and to follow the examples of ancient Greece and Rome.

In London, the print-sellers offered engravings of the classical sites, though these were frequently based more on imagination than truthful observation. If accuracy mattered, what better medium on which to base the engravings than the daguerreotype? The French had already led the way with *Excursions Daguerriennes* in which highly detailed engravings were based on daguerreotypes by Nicholas Lerebours (1807–73).

Using this as a model, Ellis planned to publish a series of sixty engravings entitled *Italy Daguerreotyped* which he believed would offer 'that verisimilitude which is required by the traveller, who wishes for reminiscences, the antiquarian who desires historical evidence of the state of buildings as existing at known times, and to all who wish to know what the buildings of Italy really are.'

He arrived in Pozzuoli on 21 May 1841 and began a photographic odyssey through Naples, Pompeii, Rome, Assisi, Pisa and Florence, arriving in Venice on 14 July. During those eight weeks he recorded 137 exposures in his notebook. It would be easy to underestimate this achievement in comparison to today's photography where labour and skill are sublimated to technology. For Ellis, the burden of the camera, lenses, metal plates and chemistry would have been considerable. As the intention was to publish accurate engraved transcriptions of the plates, size was an important factor. Smaller, more convenient plates would not provide the engraver with adequate information and detail. The practice was to have the engraving plate the same size as the daguerreotype. Despite the added inconvenience Ellis chose large 150×205 mm plates to ensure high quality results at every stage of his project.

There was a tiny body of published experience on which he could have drawn for guidance. All exposures had to be estimated by trial and error. The climate affected results. In the heat of an Italian summer, the chemicals acted variably and Ellis's patience was severely tested. A lack of understanding also led to frustration and failure. In Venice, on 17 July, he repeatedly tried to make a successful exposure. After eight attempts he put his lack of mastery down to the damp carpet in the bottom of the gondola on which the plates had been placed for safe-keeping. In Rome, he collaborated with two Italian photographers, Achille Morelli and Lorenzo Suscipi. From them he bought multi-plate panoramic sequences of the city from the Capitol Tower and San Pietro in Montorio. In all, he acquired a further forty-five plates to add to those he had made.

The scheme to publish, planned for 1 January 1845, came to nothing. It was, perhaps, a casualty of changing tastes at a time when novelty was paramount.

The surviving 159 daguerreotypes are the largest and most comprehensive group of topographic daguerreotypes and can properly claim to be the earliest photographs of Italy.   RT

Opposite: Daguerreotype of the Rialto Bridge, Venice, by Alexander John Ellis, 1841

THE ROSSE MIRROR

Even in the late twentieth century, when we are accustomed to scientific apparatus on a large scale, the Earl of Rosse's six-foot mirror is striking for its size. It is also striking for its workmanship, and when its history and context is understood it becomes a very significant object.

William Parsons (1800–67), the 3rd Earl of Rosse, had a family seat at Birr Castle in what is now County Offaly, Ireland. From the 1820s he devoted considerable time, energy and wealth to the study of astronomy. He was particularly interested in the construction of large reflecting telescopes.

In the late eighteenth century Sir William Herschel had pioneered the production of large reflecting telescopes and made many remarkable discoveries (See HERSCHEL'S SEVEN-FOOT TELESCOPE). He had turned from refractors – or telescopes using lenses – to reflectors because good lenses could not be made large enough to give the required light-gathering power. However, by the 1840s the situation had changed. There had been major developments in the production of flint glass and the design of lenses. The new observatories of the early nineteenth century were equipped with refractors which were easy to maintain and ideal for the positional astronomy currently in favour. Despite this the Earl of Rosse was determined to pursue his line of interest and continue in William Herschel's tradition.

He began by erecting a foundry and workshop in the grounds of Birr Castle, overcoming the shortage of skilled labour by training the tenants on the estate. The first mirrors were six inches in diameter, followed by fifteen, twenty-four, and eventually thirty-six inches in 1840. The three-foot mirror was a great success, encouraging Rosse to go beyond Herschel's limit of four feet and plan a six-foot reflector.

The casting, grinding and mounting of a mirror this size set a tremendous challenge to the technological capabilities of the time. The foundry had to be enlarged to accommodate the three furnaces needed to melt enough metal to fill the mould. An alloy known as speculum metal consisting of approximately two parts copper to one of tin was used. The mould was made by binding together layers of hoop iron arranged to allow air to pass through so that the metal cooled more quickly.

On the night of 18 April 1842 a crowd gathered to watch the first casting. The crucibles, holding one ton of metal each, were subjected to nineteen hours of intense heat using only turf fires. As they were poured, the crowd saw 'a burning mass of fluid matter . . . settling into a monument of man's industry for ever'. The mirror, still molten, was dragged along a railway track to the annealing oven where it was cooled gradually for sixteen weeks. The grinding and polishing could then begin. Five mirrors were made in this way, only two of which were suitable for use. The mirror in the Science Museum is the first successful one.

The mounting of the telescope proceeded throughout 1843 and 1844. The mirror had a focal length of fifty-four feet, and the enormous telescope tube was supported between two walls fifty feet high, seventy feet long and twenty-three feet apart. The tube was so large that one of Lord Rosse's friends, a certain Dean Peacock, was able to walk through it under an open umbrella. The walls ran in a north-south direction along the meridian and since the telescope had very little lateral motion it was effectively a transit telescope; that is one which views celestial objects as they pass over the meridian.

The telescope became known as the 'Leviathan of Parsonstown', and it had the distinction of being the largest in the world for seventy-five years. From 1848, the telescope was trained almost exclusively on nebulae, luminous clouds which had been seen by earlier astronomers. In 1850 the Earl of Rosse reported that through his telescope, the great nebula in the constellation Andromeda could be seen to have a spiral structure. We now know that it is a spiral galaxy like our own. He was able to see stars in so many nebulae that he became convinced that all nebulae were in fact star clusters, a theory which was later disproved.

Although the telescope was well used for thirty years there were difficulties, the principal one being the Irish weather. Many frustrating nights were spent waiting for the clouds to disperse and the constant dampness tarnished the mirror. Another was the fact that four people were needed to operate the instrument which took about ten minutes to set up and was not easy to move. A third was that the Earl of Rosse had other commitments: he was MP for the area during the difficult famine years and also President of the Royal Society (1848–54).

The endeavours of the Earl of Rosse showed the scientific community that the making of large reflectors was not impossible. When the technique of coating glass was discovered in the second half of the century reflectors gradually returned to the forefront of astronomical discovery. Many of Rosse's techniques were then copied and developed.   JAW

The 'Leviathan of Parsonstown'

ELIAS HOWE'S SEWING MACHINE

'One of the few useful things ever invented' was Mahatma Gandhi's assessment of the sewing machine. Up to 1845, there had been several aspirants to the title of 'inventor', yet each machine was flawed in some way. In that year, Elias Howe Jr. (1819–67) triumphed in incorporating the most significant ideas to date into a machine which became a starting point for the sewing machine and mass-produced clothing industries. Born in Spencer, Massachusetts, Howe was the son of an English immigrant farmer and succeeded where others had failed, not only through ingenuity, but because his work was timely. It slotted neatly into a period when machine-made cloth had become commonly available and engineering had advanced sufficiently to allow the manufacture of affordable machines of such complexity.

Rather than face the mechanical nightmare which imitating manual sewing would have incurred, Howe neatly bypassed the problem by using the 'eye-pointed' needle and the lockstitch. This needle was devised by the Londoner C F Weisenthal, in 1755, for embroidery. The lockstitch combined two separate threads, one each side of the fabric; it was developed in the 1830s by one Walter Hunt of New York. Luckily for Howe, Hunt did not patent the lockstitch, so anyone could use his idea.

In the Howe sewing machine, an eye-pointed needle penetrates the cloth only just far enough to pass the thread through. On retracting, the friction of the cloth on the cotton thread causes only a small loop to be left on the other side of the cloth, just large enough for the second thread to be passed through it. Howe used a curved needle, swinging sideways in an arc. The secondary thread was carried in a shuttle; in this way, Howe had applied one of the principles of weaving to seam sewing. Unlike modern machines, the cloth in Howe's machine hung vertically from pins along the edge of a plate which edged forward, stitch by stitch. After a few inches' sewing, the plate had to be reset and the garment re-pinned in a new position.

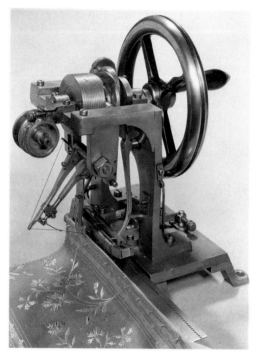

The Howe sewing machine, showing cloth pinned up for sewing

Having been granted a United States Patent in 1846, Howe looked to England for a wider market for his contrivance. A machine was taken to England, where William Thomas of Cheapside, an umbrella and corset maker, became interested in its promise of greater productivity. An agreement whereby Thomas patented the machine in England and, in return, paid Howe royalties was concluded. But the English venture turned out to be a disaster. The arrangement with Thomas did not work out (for Howe, at least) and Howe was beset by debt and poverty. His wife became ill with tuberculosis, from which she died in 1849. Back in America the same year, he found that the technology and application of sewing machines had advanced; he also felt other sewing machine makers had been using his patented ideas.

He embarked on a series of legal battles, challeng-

ing makers whose machines incorporated features which he claimed as his own; he even took on the formidable Isaac Singer (1811–75) and won. Patent battles tied up the whole sewing machine industry, and progress ground to a halt. But by 1856, Howe had been persuaded that peace rather than conflict was the way ahead and he accepted an agreement whereby all significant patents would be pooled for the benefit of all concerned. Howe's fortune was, at last, assured; it is estimated that, by 1867 when his patent expired, he had received two million dollars. Sadly, he did not live to enjoy his fortune.

The machine in the Science Museum has direct links with his ill-fated London venture. It was given in 1919 by the son of William Thomas, who stated that it was the machine brought from America to England in November 1846.

The growth of the sewing machine business, of which Howe laid one of the foundations, turned out to have profound industrial and social implications. The labour-saving possibilities alone were enormous; in 1861, a shirt took just over an hour to make by machine, compared with fourteen hours by hand. Mechanized sewing spread to many products other than clothing: ships' sails, cloth bags and sacks, and boots and shoes. In the latter case, the power-driven needle was as welcome as mechanization itself. 'New technology' disputes abounded and manufacturers often resorted to relocation to capture more compliant workers.

The acceptance of sewing machines into the home occurred in the United States before it did in England. Singer, in particular, spotted the possibility of selling domestic machines in huge quantities; his marketing was aggressive, and he pioneered hire-purchase schemes in order to penetrate the poorer end of the market. As with so many labour-saving contrivances, the take-up in England occurred some years later, the reason being that people wealthy enough to buy machines could employ servants. In contrast, the United States enjoyed fuller employment and perhaps more democratic attitudes.   ED

JOULE'S PADDLE-WHEEL APPARATUS

James Prescott Joule (1818–89) was born in Salford, a part of the Manchester conurbation then the centre of cotton production in Britain. The cotton industry relied on steam power and Joule's work laid one of the foundations of the science of thermodynamics, which provided a scientific basis for the development of the steam engine and other heat engines.

Joule's father was a wealthy brewer, and for much of his life Joule was able to follow his interest in science in a private laboratory. His earliest researches were into electromagnetic engines, the primitive electric motors that seemed to promise an inexhaustible supply of motive power. Joule soon showed that an electromagnetic engine was much more expensive to run than a steam engine and began instead to investigate the heat produced by an electric current, quickly discovering the law relating the quantity of heat produced to the current and to the resistance of the conductor in which it flows. This was a most significant discovery, but Joule went further, and in doing so began to challenge the accepted theory of heat.

There were two rival theories of the nature of heat: the 'caloric' theory, that it was a subtle fluid, and the 'dynamical' theory, that it represented the motion of the atoms of matter. Either theory could explain known thermal phenomena with varying degrees of difficulty. In both theories it was assumed that the total heat of a system of bodies was conserved. Thus, for example, when two objects were rubbed together it was believed that they became hot because heat was transferred from somewhere else in the system, not because the friction actually generated the heat.

Joule performed a series of electrical experiments in which he first proved that heat was produced by an electric current, not merely transferred from another part of the electrical system which would have had to become cooler. Then he showed indirectly that heat could be converted into mechanical work. Finally he measured the mechanical work done in producing a unit of heat.

Joule reported his results at the meeting of the British Association in 1843, but they aroused no interest. The idea that heat was conserved was too strongly established to be shaken by a little-known person working outside established scientific institutions, and many questioned the reliability of Joule's conclusions, drawn from a few measurements of small temperature changes.

Undaunted, Joule built new apparatus in which mechanical work was converted into heat directly, by stirring water vigorously with a horizontally rotating paddle wheel. The work needed to rotate the paddle wheel was provided (as in his previous experiments) by weights falling through a measured distance. Joule reported his new measurements to the British Association meeting in 1845, again with no obvious impact, and yet further measurements to the meeting in 1847. This time his ideas created much more interest, largely because they attracted the attention of the young William Thomson (1824–1907), who, as Lord Kelvin, was to become one of the greatest physicists of the age.

Thomson could not immediately accept Joule's work, but in the end it led him, along with men like Clausius and Rankine, to develop the theory of thermodynamics. Joule's acceptance was complete when the Royal Society published his account of new and more accurate experiments in 1850.

Joule did not have the mathematical skill to contribute further to thermodynamics. He had been fortunate in meeting William Thomson. J R Mayer, in Germany, had proposed similar ideas in 1842, and measured the mechanical value of a unit of heat, though less rigorously than Joule. His work was mostly ignored.

Joule's research established the first law of thermodynamics, the principle of the conservation of energy, though it was the German physicist Hermann Helmholtz (1821–94) who formulated this general conclusion, in 1847. Energy is conserved, not heat: heat is one form of energy and may be converted into another. Joule's work extended the dynamical theory of heat, which he had always supported, and demolished the caloric theory, which became untenable once it was known that heat is not conserved.

The 'mechanical equivalent of heat' remained important for many years, featuring strongly in school physics. Not until the 1960s and the introduction of the International System of Units was Joule's work taken to its logical conclusion. The mechanical equivalent of heat is now a historical curiosity because all types of energy are measured by the same unit. That unit, first defined in 1889, is called the joule.   CNB

ORIGINAL MAUVE DYE

If your discovery does not make the goods too expensive, it is decidedly one of the most valuable that has come out for a very long time. This colour is one which has been very much wanted in all classes of goods and could not be obtained fast on Silks, and only at great expense on cotton yarns.

So wrote Robert Pullar of John Pullar and Son, a leading firm of Scottish dyers, on 12 June 1856 following evaluation of fabric samples treated with mauve dye developed by W H Perkin. Some eighteen months later this novel chemical product was being despatched from the Greenford Green factory of Perkin and Sons, having been bought by the largest silk dye works in London, an event now reckoned a milestone in nineteenth-century technological progress. Not only was this the first step in the industrialization of organic chemistry but also the beginning of the commercialization of scientific invention.

Before this time dyestuffs were virtually all of vegetable or animal origin and their variety had not increased since the Middle Ages, despite intense efforts at improving methods of dyeing in the wake of the vast growth of textile manufacture during the Industrial Revolution. The initial context of Perkin's discovery was however quite unrelated to these efforts. William Henry Perkin (1838–1907), the son of a builder, showed a keen interest in chemistry from an early age and enrolled at the Royal College of Chemistry in 1853 where he attended the lectures of the German chemist A W Hofmann (1818–92), renowned for his research and teaching abilities.

By the middle of the nineteenth century quinine was much in demand for the fight against malaria, but it was expensive. As a result Hofmann and a number of other chemists were investigating alternatives. From their very limited knowledge of the relationship between chemical composition and molecular structure it was not unreasonable for Hofmann to speculate in 1849 that quinine might be

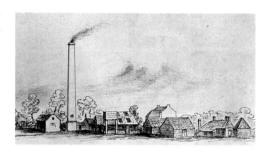

Perkin's factory at Greenford Green near London

synthesized by the addition of water to naphthylamine, a substance obtainable from coal tar.

In 1856 Hofmann's young research assistant, Perkin, resolved to attempt a synthesis based upon the addition of oxygen rather than water, selecting as his starting material allyltoluidine, which can also be obtained from coal tar. Perkin undertook the oxidation at home, having in his enthusiasm for research equipped part of a room in his father's house for this purpose soon after beginning his studies under Hofmann. Working during the Easter vacation of 1856 his oxidation yielded none of the desired colourless quinine but only a dirty reddish-brown sludge. Accordingly he repeated the oxidation with the simplest base available from coal tar, aniline. The product this time was a black precipitate, which upon drying and treatment with methylated spirit, yielded an intense purple solution. Perkin found to his surprise that it dyed silk a beautiful colour and better resisted the fading effects of light than existing dyes used at that time.

The eighteen year old pursued his discovery with the vigour of youth and the wise counsel of those in the dyeing industry with whom he was put in contact. A provisional patent was deposited in London by 26 August of the same year and the final version sealed the following 20 February. His resignation from his post at the Royal College of Chemistry in October 1856, which Hofmann regarded as

foolhardy, further testified to Perkin's commitment to make money from his synthetic dye. To produce the dye on a commercial scale required Perkin's father to recognize that his son's interest in chemistry had not been in vain, as he once had feared, and that the necessary capital expenditure would be a sound family investment. This proved to be a wise judgement.

Perkin's discovery gave impetus to a new coal tar dyestuffs industry in which the level of protection afforded by patents was much less than it is today. Fortunately whilst slight modifications of the original process became rife, none outdid the economy of the original method. In addition Perkin kept up his research activities and introduced new colouring materials himself, notably 'Britannia Violet' in 1864, derived from magenta. This helped to keep the Greenford factory operating at a profit as more brilliant dyes displaced mauve from the market after a span of less than ten years.

In 1869 Perkin devised two new methods which allowed the economic manufacture of alizarin, the natural colouring matter of madder, the prime red dye of the period, the synthesis of which had been reported by Graebe and Liebermann in 1868 but by a process too expensive to be of commercial interest. By the end of 1869 Perkin's company had made its first ton of alizarin, expanding output to over 200 tons per annum by 1871.

Perkin had however always hoped to devote himself completely to pure research and by 1873, at the age of thirty-five, he found that his factory and patents could guarantee his 'retirement'. He sold his interests the following year. This early work was the foundation of the artificial dyestuffs industry we know today. However, British industrialists failed to build on Perkin's work. As a result, Germany, whose scientists were in the forefront of new branches of chemistry, soon took the lead in the production of artificial dyes. DAR

LEWIS CARROLL'S PHOTOGRAPHS

Early in 1855, when photography was not yet twenty years old, an Oxford student, Charles Lutwidge Dodgson (1832–98), son of the rector of the Yorkshire village of Croft, went to the nearby town of Ripon and 'Got my likeness photographed by Booth. After three failures, he produced a tolerably good likeness, which half the family pronounce the best possible, and half the worst possible'.

Dodgson is today best known by the pen-name 'Lewis Carroll', under which he wrote two of the most enduring children's books, *Alice's Adventures in Wonderland* and *Alice Through the Looking-Glass*. His Ripon portrait was taken with the wet collodion process which its inventor, Frederick Scott Archer, did not patent, but made freely available to all. Unlike the earlier calotype and daguerreotype, therefore, it had made photography relatively cheap, reliable and accessible.

One of Carroll's uncles, Skeffington Lutwidge, was an amateur practitioner. Later in 1855, when Carroll was on summer vacation from Christ Church, Oxford, he observed his uncle's variable results and wrote *Photography Extraordinary*, the first of his many works to use the medium as subject matter. It playfully proposed an imaginary camera which would turn a writer's half-formed ideas into a finished novel.

Back in Oxford, after taking advice from Skeffington, he and a fellow student, Reginald Southey, another enthusiast, travelled to London in March 1856 to buy a camera at Ottewill's, Islington.

Though the wet collodion process was easier to manipulate than the daguerreotype or calotype, to make a negative required photographers to mix, precisely and dextrously, their own chemicals (in the dark) and to spread them evenly over a sheet of glass. To do this carelessly, or leave any part of the glass uncovered, resulted in a faulty picture. The negative had to be placed in the camera, and the picture taken, while the chemicals were still moist (hence 'wet' collodion): when dry, they lost their photosensitivity. Not surprisingly, there were many failures.

Lewis Carroll: self-portrait
probably taken at his rooms, Christ Church, Oxford.
Opposite: *Reginald Southey and Skeletons, 1857*

After photographing buildings, statues, medical specimens, trees, gardens, and still lifes, Carroll turned to portraiture. He began a remarkable series of university colleagues and friends, working in his rooms in Christ Church or outside in the college gardens. Reginald Southey was clearly posed outdoors, apparently in front of a sheet of plain wood erected for the purpose. He was a medical student, and the skulls and skeletons with which he is standing suggest that he had views on Darwin's theory of evolution, the subject of furious debate in Oxford at the time.

Later, Carroll had a studio, or 'glass house', built on the roof of the main ('Tom') quadrangle, to which he brought hundreds of sitters. For twenty-five years he seems to have spent more time on photography than on writing or on his work as a mathematics tutor.

Increasingly, Carroll's subjects were young girls. Alongside his diary entries for March 1863, he listed 103 whom he wanted to photograph, in alphabetical order of their first names – he must have known them all intimately, at least in his imagination. It is a real achievement to have persuaded those who did pose for him to keep so still in front of his camera. The shortest exposure time he recorded is forty seconds, and his diary is as full of notes of failed photographs as Victorian family albums are full of blurred faces, hands and eyes. But Carroll's success made him, as the distinguished photo-historian Helmut Gernsheim has said, 'The greatest Victorian photographer of children'.

Though Carroll's pictures, of adults and children, are less stilted and artificial than many in the Victorian era, they are nonetheless carefully considered. Poses and grouping were his 'favourite branch of the subject' and he concentrated particularly on 'the arrangement of the hands'. Because of his long exposure times, sitters are virtually never shown smiling, but *Reginald Southey and Skeletons* has something of the humour found in Carroll's books.

The mystery about Lewis Carroll's photography is that he abandoned it in July 1880, without warning or explanation. There are several possible reasons. One is that he disliked the dry plates which were replacing wet collodion. Another is that his enthusiasm for photographing young girls, preferably in very few clothes, finally got him into trouble. But the most convincing one is somewhat less sensational.

As the sale of Carroll's books grew, he may well have taken the decision to concentrate on writing. A year later, he resigned his mathematics lectureship: 'I shall now have my whole time at my disposal and, if God gives me life and continued health and strength, may hope, before my powers fail, to do some worthy work in writing'. As he approached fifty, he turned to his real vocation as one of the greatest humorists in English literature.   CJF

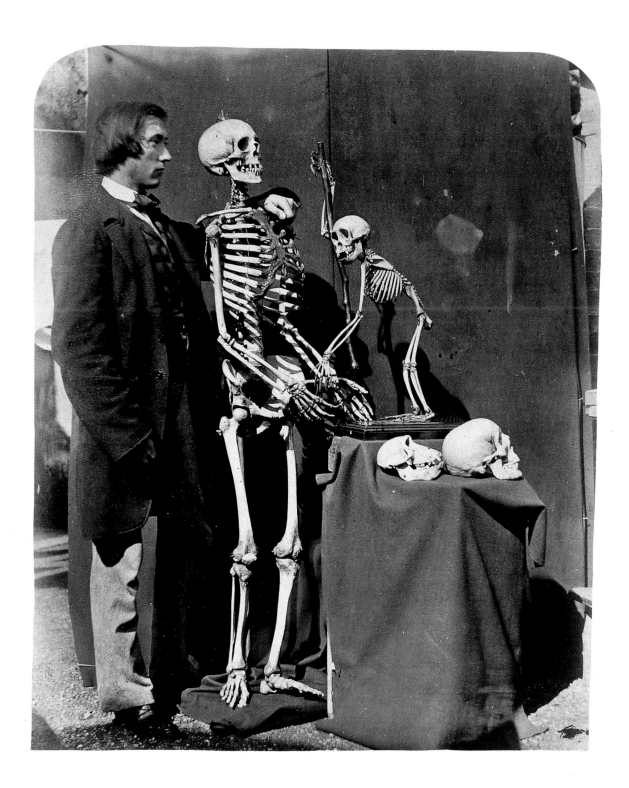

THOMSON'S MIRROR GALVANOMETER

Thomson's mirror galvanometer was the only instrument which could reliably detect signals sent through the first Atlantic telegraph cable, laid in 1858. The instrument shown here was used at the Newfoundland end of the cable.

The galvanometer, an instrument for detecting and measuring electric currents, was developed by the Italian scientist Leopoldo Nobili (1784–1835) in the 1820s. It was the first practical application of the discovery, announced by Oersted in 1820, that an electric current in a wire produces a magnetic field. In a galvanometer the current being measured is passed through a coil, creating a magnetic field. The magnetic field interacts with a magnet, which can turn (in some later instruments the magnet is fixed and the coil turns). The amount the magnet turns is a measure of the current flowing in the coil. Usually the magnet carries a pointer which moves across a scale, and in some instruments the magnet and the pointer are the same piece of iron.

The electric telegraph began to be exploited commercially in the 1840s. Nearly all telegraphs used receiving instruments which were essentially galvanometers. When an electric current was sent through the telegraph wire the galvanometer needle at the other end moved in response. In most telegraphs messages were sent by sequences of long and short current pulses, using Morse or similar code.

Most inland telegraph wires were carried on poles in the open air, although underground cables were sometimes used for short distances in cities. To establish telegraph connections with other countries it was necessary to adopt submarine cables. Submarine cable technology was pioneered by British scientists and engineers. By the 1870s it had created a forerunner of today's 'global village', reducing the time for messages to travel from continent to continent from days or months to hours.

Long submarine cables have very different electrical characteristics to wires carried on poles and spaced apart in the air and as a practical consequence signalling becomes very slow. In 1858, the electrician

Sir William Thomson, Lord Kelvin

to the Atlantic Telegraph Company, Wildeman Whitehouse, attempted to solve the problem by using very high voltages from an induction coil to send the signals. The result, however, was damage to the cable.

The problem was solved by William Thomson, (1824–1907) the Professor of Natural Philosophy at Glasgow University who was already well known for his electrical work. He appreciated that the solution to the problem of sending a detectable signal through the cable was to use a small current and a very sensitive galvanometer as current detector. With a small current the charging time was reduced but the problem of detecting the current was greatly increased.

Thomson's galvanometer operated in the same way as all other galvanometers, but its construction was very light. The magnet was small, and instead of having a pointer it carried a tiny mirror, fixed to the magnet. A beam of light shone on the mirror and the

slightest movement of the magnet and mirror would cause the reflected beam of light to move across a scale. The system could detect very much smaller currents than any previous meter, and made possible the first Atlantic telegraph in 1858.

Thomson was knighted for his contributions to the electric telegraph, and subsequently raised to the peerage as Baron Kelvin of Largs, the first person to be so honoured for his contributions to science.

The 1858 cable only lasted a short time (there had been manufacturing and cable-laying problems as well as the damage caused by Whitehouse's experiments). When a more successful cable was laid in 1866 the electrician Latimer Clark arranged a striking demonstration in which a simple electric cell consisting of a silver thimble, a small piece of zinc, and acid, sent a current across the Atlantic and produced a strong deflection in a mirror galvanometer.

Two people were needed to receive Morse code signal by means of the mirror galvanometer: one watched the beam of light, noting the Morse dots and dashes, and calling out the letters for the second to write down. By 1867, however, Thomson had developed an improved instrument for receiving signals, his 'siphon recorder'. This was still basically a galvanometer, but instead of carrying a mirror the magnet carried a small, siphon-shaped glass tube which had one end in a vessel of ink and the other end close to a moving strip of paper. As signals were received the end of the siphon tube was moved sideways across a strip of paper which was drawn through the receiver. The siphon recorder was also a very sensitive instrument, and since it made a permanent record on the paper of the messages received it was no longer necessary for the instrument to be watched continuously.

Although mirror galvanometers were soon replaced as telegraph instruments by Thomson's siphon recorder, they continued to be used for more than a century when electrical measurements at high sensitivity were required. Electronic, digital meters are now taking their place.   BPB

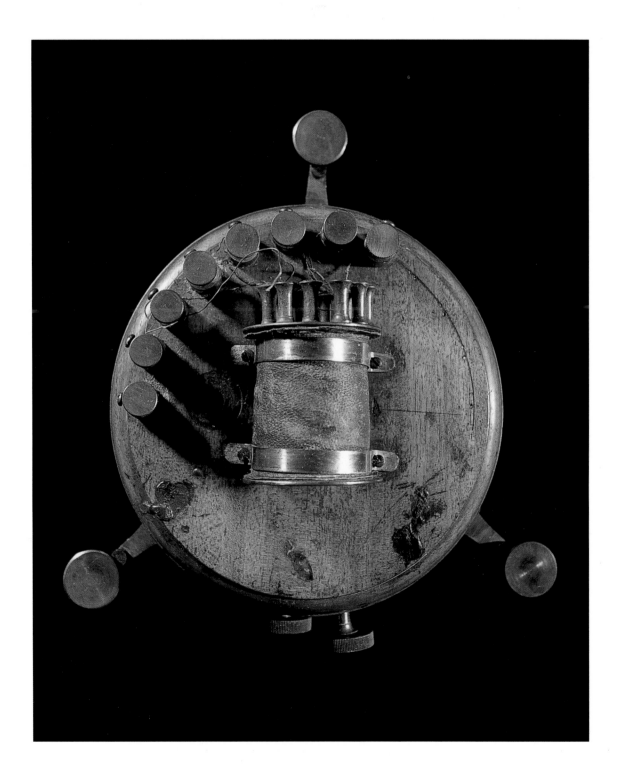

THE KEW PHOTOHELIOGRAPH

The Sun is not the perfect body that Aristotle would have had us believe; it is occasionally flawed by the presence of spots. It is curious that although sunspots were observed since the early days of the telescope it was not until over two hundred years later that the German amateur astronomer Samuel Heinrich Schwabe (1789–1875) discovered that their occurrence is cyclic. They wax and wane over a period of about eleven years although later research has identified longer cycles as well.

It was only natural that British astronomers should want to conduct regular solar observations to confirm and extend Schwabe's findings. Sir John Herschel (1792–1871) was particularly vocal in demanding a suitable instrument for this work, and in 1854 he was able to report to the Kew Committee of the British Association that £150 had been vouchsafed by the Donation Fund of the Royal Society for the purpose.

The King's Observatory at Kew had been built so that George III could observe the Transit of Venus in 1769, but the British Association took it on in 1842, in part as 'a repository and station for trial of new instruments'. To begin with, these were primarily meteorological and magnetic instruments.

As a pioneer of astronomical photography with a technique using wet collodion plates, Warren De La Rue (1815–89) was a natural choice to superintend the construction and use of the new solar instrument. The Kew photoheliograph was commissioned from the optical instrument maker Andrew Ross: a combined camera and telescope with a three and a half-inch lens of fifty inches focal length, a spring-loaded shutter to permit rapid photographs of the Sun and a clockwork drive to compensate for the Earth's rotation. The Kew photoheliograph was first used in 1858, in time to make routine observations around the minimum of the solar cycle. Only two years later, it was dismantled to participate in an eclipse exhibition of historic significance.

It can have been no mean feat to transport the photoheliograph, plus its cast-iron mounting and sundry photographic apparatus, by sail and stage-coach to a village in northern Spain. But on two counts De La Rue was convinced that it would be worthwhile. First, the solar eclipse of 18 July 1860 would be the first opportunity to test wet collodion plate photography at a solar eclipse (earlier photographic attempts had produced meagre results). Second, De La Rue would seize the opportunity to settle a scientific question by photographic means.

The question revolved around the nature of the controversial plumes seen on the limb of the Sun in eclipse, when its blinding light is obscured momentarily by the Moon. Did these 'prominences' arise in the Earth's atmosphere or in the Sun's? A relatively simple way to settle the matter would be to record the eclipsed Sun from different locations. If prominences were some new meteorological phenomenon, the observations from widely separated sites ought to differ; but if prominences were solar in origin, they should match.

De La Rue established his observatory on a threshing floor in Rivabellosa, on the borders of Rioja country. Meanwhile, Father Angelo Secchi (1818–78) set up his telescope at Disierto de las Palmas, 500 kilometres south-east by the Spanish coast. De La Rue was able to take thirty-five usable photographs, including two crucial ones during the precious minutes of total eclipse.

These photographic plates, wrote De La Rue, 'depicted the luminous prominences with a precision as to contour and position impossible of attainment by eye observations'. Although the faint solar corona visible during the eclipse eluded photographic record, a plethora of prominences appeared. When De La Rue subsequently compared Father Secchi's photographs with his own, he was thrilled to find that they coincided. There was no longer any question: prominences belong to the Sun itself.

Back in England, the photoheliograph provided a series of photographs covering an entire solar cycle. It took its last solar photographs at Kew at the end of March 1872.   JD

The temporary observatory at Rivabellosa. The lens of the photoheliograph is just visible above its shelter.
Warren De La Rue appears twice; this view was montaged from two separate photographs

THE FIRST PLASTIC

Plastics are now very familiar to us. They can be used for an almost infinite variety of purposes, from fabrics, furniture and crockery to cars, aeroplanes and space shuttles. What is generally accepted as the first plastic, Parkesine, was formulated about 130 years ago.

A plastic is a material that can be moulded or shaped into different forms under pressure and/or heat. Chemically, plastics consist of long chains of molecules made up predominantly of carbon and hydrogen atoms. Their great advantage over other materials is that today plastics can be designed for almost any purpose.

The earliest plastics, beginning with Parkesine, were semi-synthetic, being based on natural substances like cellulose obtained from sources such as cotton flock or wood fibres. Parkesine dates back to the middle of the nineteenth century. This was a time when much work was being carried out to find materials which would make satisfactory replacements for materials such as ivory, amber and tortoiseshell which were becoming increasingly rare and expensive.

Parkesine was the invention of an energetic Birmingham scientist, Alexander Parkes (1813–90). He extended the work of a Swiss chemist, Schönbein, who had produced pyroxylin (cellulose nitrate) in 1845 from cellulose fibres mixed with nitric acid. Parkes discovered that if he mixed cellulose nitrate with plasticisers and solvents, he produced a mouldable substance, which he called Parkesine. He exhibited it in a variety of forms at the International Exhibition of 1862, including combs, hair slides, billiard balls and carved plaques. Parkes' product impressed the judges at the exhibition and he was awarded a bronze medal for 'excellence of product'.

Alexander Parkes went on to produce a wide range of objects in Parkesine. They show a variety of uses as well as rich jewel-like colours, and include crude samples, fishing-rod spools and a miniature, intricately carved head of Christ crowned with thorns, in imitation amber.

Alexander Parkes:
lithograph by A Wivell junior, 1848

The success of Parkesine at the 1862 International Exhibition, and the fact that he believed that he could reduce the production price to one shilling (5p) a pound, encouraged Parkes to establish the Parkesine Company Limited in 1866 with a capital sum of £10,000. Although set up with enthusiasm, the company soon foundered and went into liquidation in 1868. Parkes himself blamed the flammability of the material. Others have suggested that the Parkesine Company failed due to Parkes' obsession with keeping the price low, and hence, perhaps, using materials which were cheap and of poor quality. Maybe he would have succeeded if he had marketed Parkesine as a luxury product at a higher price. Parkes' work formed the foundation for the future development of Celluloid.

Daniel Spill (1832–1877), who had been Parkes' Works' Manager, then attempted to make a commercial success out of Parkesine, which he renamed Xylonite. Spill set up the Xylonite Company in 1869 to produce and market this product. At almost the same time, parallel work was being carried out in the United States by John Wesley Hyatt. The earliest cellulosic plastics (both in the USA and Britain) had been greatly plagued by the problem of flammability. Factory fires were a continuing problem with plastics based on cellulose nitrate, never to be totally resolved.

Hyatt realized that camphor made an excellent plasticiser for cellulose nitrate, producing a more stable and usable product. This he christened Celluloid. One of the initial uses of Celluloid was to make dental plates (replacing their unsatisfactory rubber-based predecessors), and in 1870 Hyatt and his brother Isaiah set up the Albany Dental Plate Company. By 1872 the name had been altered to the Celluloid Manufacturing Company. The Hyatt brothers' company was very successful, producing large numbers of saleable goods, including combs and cosmetic boxes.

Spill felt that Hyatt's work on Celluloid infringed the British patents, and entered an eight-year-long litigation battle in the American courts. After an initial favourable verdict, judgement eventually went in favour of Hyatt, and Celluloid became a runaway commercial success. Spill died an embittered man.

So, finally, how should we assess the contribution of Parkesine and the work of Alexander Parkes to the history of plastics? Parkes was the first to produce a mouldable material based on cellulose nitrate, which later developed into what we know as Celluloid, abundant in the late nineteenth and early twentieth centuries for hair adornments, for decorative items such as dressing-table sets, even bags and, perhaps most notably, low-cost, wipe-clean collar and cuffs. His work provided an impetus for what is now a vast industry of many plastics used for every conceivable purpose. Parkes can be seen to have earned his title as the 'Father of Plastics'. SM

Comments from Susan Cackett and Colin Williamson

THE LENOIR GAS ENGINE

In *The Engineer* of 26 October 1860, there appeared a brief notice headed 'New Motive Power – Lenoir's Gas Engine'. It stated:

A machine placed in a corner of a room, of a size proportionate to the force intended to be developed, the pipe which brings into the manufactory the gas which lights it, a little electric pile on a bracket against the wall, a little water supplied by the water company and atmospheric air, and we have all that constitutes the moving power of M. Lenoir. The machine itself is little more than an ordinary horizontal engine, but instead of it being brought into operation by the expansive force of steam, motion is produced by the combustion of a mixture of gas and atmospheric air; and this combustion is brought about by the action of an electric battery.

The Lenoir gas engine was the first practical internal combustion engine, and a forerunner of today's petrol and diesel engines. Piped supplies of coal gas, principally for lighting, had become widely available in large towns by the middle of the nineteenth century and devices to produce an electric spark had been developed. A compact engine which used gas and an electric spark to ignite it was therefore seen as a major improvement on the steam engine, which needed a boiler, bulky fuel and a large water supply.

The cannon, of which the first recorded use is believed to have been in the fourteenth century, may be regarded as the ancestor of the internal combustion engine, so the idea of producing mechanical work from the combustion of a rapidly burning fuel in an enclosed space has a long history. Late in the seventeenth century, the physicist Christiaan Huygens (1629–95), noted for his wave theory of light, made the first gunpowder engine which could actually work. In Huygens' machine, a charge of gunpowder was fired in the bottom of a cylinder in which a piston rested at the top; valves allowed much of the expanding gases to escape; as what remained cooled, a partial vacuum was formed in the cylinder causing the piston to be forced downwards by atmospheric pressure. A weight, attached to the piston by a rope passing over a pulley, was thereby raised. There is clearly a parallel here with the early steam engine whereby atmospheric pressure acted on a piston in a cylinder in which a partial vacuum was created by the condensation of steam.

From the 1790s onwards, experimenters made engines using various types of fuel, including hydrogen and other gases obtained from the distillation of coal, wood or oil. They also tackled the problem of producing continuous operation, of which Huygens' engine was clearly not capable. Some of the engines developed were direct acting and others worked on the atmospheric principle. Lenoir's machine was the first to progress beyond the experimental stage into commercial production. It is thought that the total number built approached 500; most were made in France but some were manufactured in England, Germany and North America.

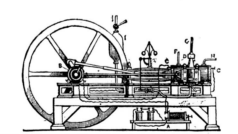

The Lenoir Gas Engine.—A Number of these useful and economical Engines to be seen in actual work in town and country. Being free from all danger, they can be used where steam power is totally inadmissible. — Exhibition and every information supplied at 40, Cranbourne-street, Leicester-square, W.C. (I 681)

An advertisement for the Lenoir Gas Engine, from *The Engineer*, 22 December 1865

Production ceased around 1869.

Jean Joseph Etienne Lenoir (1822–1900) was a Belgian who went to Paris at the age of twenty-six to work as a metal enameller. His inventions included a method of electroplating and a railway telegraph system. His gas engine resembles in form and sequence of operation a single-cylinder double-acting steam engine, in that pressure acts alternately on each side of the piston. Air and gas are drawn into the cylinder during the first part of each stroke. The mixture is then fired by an electric spark and the expanding hot gases drive the piston to the end of the stroke. As the piston moves to the end of its travel it forces out, through the exhaust valve at the other end of the cylinder, the spent gases from the previous power stroke. Cooling of the cylinder is assisted by a water jacket. A battery formed from two Bunsen cells provides the ignition via an induction coil and a distributor serving a sparking plug at each end of the cylinder.

About a hundred Lenoir engines were built in England by Reading Ironworks Limited. The engine in the Science Museum is nominally of half horsepower, and drove the workshops of the Patent Office Museum from 1865 to 1868. An engine of this size would have cost £65. With regular skilled maintenance, it gave satisfactory service. A contemporary record states that a peak indicated output of about one horsepower (0.75kW) was achieved when running at 110 revolutions per minute. However in less skilled hands Lenoir's engines were often troublesome and their performance was soon surpassed by others – notably those developed by the German partnership of Otto & Langen. To Lenoir, though, must go the credit for pioneering work on which others were able to build.   PDS

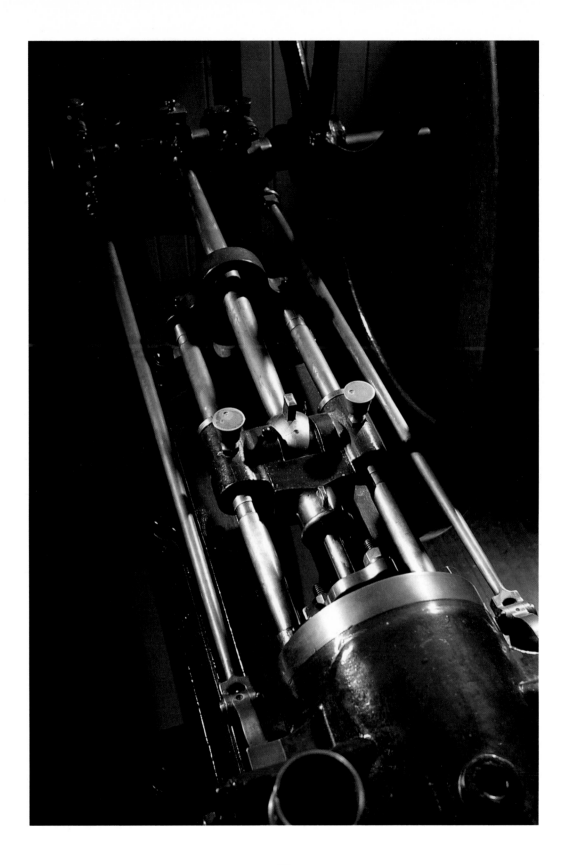

BESSEMER CONVERTER

The Bessemer Converter was developed in 1856 by Henry (later Sir Henry) Bessemer (1813–98) for the manufacture of bulk steel.

Steel is simply iron containing a small amount of carbon (typically 0.1 to 1.5 per cent depending on the properties required) to give a versatile material suitable for many applications. Today a wide range of steels and steel alloys can be produced, but prior to Bessemer's invention, steel was only available in small quantities of variable quality produced by either the cementation or crucible processes. Engineers were restricted to the use of wrought iron, which was relatively soft and weak, or cast iron which was hard but brittle.

Bessemer is often described as a 'professional inventor' due to the variety of products he devised. He was not a metallurgist, but was trying to find a malleable form of iron which would be tough enough to make shells for a rifled gun barrel he had invented. His small workshop at Baxter House, Paddington was to become the focus of his activities.

After some preliminary experiments, Bessemer realized that if sufficient air could be blown into molten pig iron (which contains about three per cent carbon) some of the carbon would combine with the oxygen in the air and be burnt out. Once the iron was melted and poured into the vessel, no further fuel would be required as sufficient heat would be obtained from the chemical reaction.

In fact the reaction was extremely violent and during his first experiment no one could approach the converter to turn off the blast, so they just had to watch from a respectful distance until things had subsided. Bessemer made some modifications to the design of the converter to contain the burning debris and in August 1856 he invited the engineer, George Rennie (1791–1866), to witness his new process.

Rennie was immediately enthusiastic and arranged for Bessemer to present a paper at a meeting of the British Association for the Advancement of Science at Cheltenham a few days later. Ironmasters present were very sceptical of a paper entitled 'The Manufacture of Malleable Iron without Fuel', especially as it was to be presented by an outsider.

However, several leading ironmasters approached Bessemer for patent rights and visited Baxter House for demonstrations. Having paid considerable sums of money in royalties, they went away to conduct experiments of their own. Without exception, these were unsuccessful. Naturally having parted with their money, the ironmasters rounded on Bessemer, but neither he nor anyone else could offer a solution at the time. Later it became evident that the problem was caused by phosphorus in the iron ore which made the resulting steel brittle and useless. By chance, Bessemer had obtained iron made from one of the few sources of low phosphorus ore in Britain, Blaenavon in South Wales.

Other low phosphorus iron ore deposits occur in the Lake District. Here the Barrow Haematite Ironworks began experiments in 1864 using the small converter now in the Science Museum. The first cast of molten steel was poured in 1865. Having proved successful in trials, the small converter was removed in 1866 and a full-size Bessemer works established. By the next year, the company was the largest Bessemer steelworks in Britain. From his first experiments, Bessemer had modified the design to a tilting converter, which allowed the vessel to be turned on its side to pour out the molten steel.

Gradually the Bessemer process became established using low phosphorus ores, but efforts to solve the phosphorus problem were not a success. The eventual solution was devised by two cousins, Sidney Gilchrist Thomas (1850–85) and Percy Carlisle Gilchrist (1851–1935). Thomas had been challenged by a chance remark at an evening class that whoever solved the problem would make a fortune. After some preliminary experiments (initially carried out in his bedroom) he enlisted the help of his cousin, a steelworks chemist. They worked out that if the converter was lined with bricks which were chemically basic (a base is able to neutralize acid), the furnace lining would neutralize the acid phosphorus and remove it from the reaction. They had great difficulty in convincing ironmasters that they had succeeded where others had failed, but it provided the solution and was adopted from about 1880.

In the meantime, another furnace had been devised by Sir William Siemens (1823–83) working in Britain and, independently, by Emile Martin (1824–1915) in Sireuil, France. Their 'open hearth' furnace could be controlled more easily to get the carbon content of the steel exactly right and could also recycle steel scrap. From the early 1870s this process developed alongside Bessemer's, but was also dependent upon the results of the Gilchrist and Thomas work for its ultimate success.

The availability of large quantities of steel at an acceptable quality and price added a new dimension to engineering. It made possible much larger bridge spans and ships, taller buildings, stronger and lighter machinery, and steel rails that paved the way for heavier and faster trains.

In more recent times, bulk steel making has relied entirely upon a derivative of the Bessemer process, Basic Oxygen Steelmaking (BOS). A much larger converter is used and pure oxygen is pumped into the molten metal using a water cooled lance inserted in the mouth of the vessel. One converter can produce over 350 tonnes of steel in forty minutes.   SJC

'SWAN' AND 'RAVEN'

In June 1867 the retired engineer William Froude (1810–79), assisted by his son Edmund (1846–1924), began testing a series of model ship-hull shapes in a small creek off the River Dart opposite the town of Dartmouth in Devon. These models, which Froude named *Swan* and *Raven*, had as their principal purpose the establishment of a method whereby a scale model could be used to predict the resistance or drag experienced by a full-sized ship as it moved through the water. Such information was necessary to enable naval architects to specify the size of steam engines needed to drive a ship through the water at its required speed.

Although a great deal of effort had been expended in the eighteenth and early nineteenth centuries in using models to estimate the resistance properties of various hull shapes, all experiments had failed to achieve their purpose. The reason for this was that nobody had been able to discover how model results could be reliably scaled up. It was well known that they did not scale up directly in proportion to the scaling factor of the model. The successful application of ship models to ship design therefore needed a law to relate the model's resistance, and the speed at which it was measured, to the full-sized counterpart. The law the Froudes were trying to establish was known as the law of similarity.

To investigate and discover the law of similarity William Froude constructed three sets of *Swan* and *Raven* models. The first were three feet in length, the second six feet and the third twelve feet, thus introducing a scale factor of two between successive models. Each set of models was then towed simultaneously by a steam launch, their respective resistances being measured by means of a recording dynamometer which also recorded the launch's speed. This procedure was carried out at a number of different speeds and the resulting speed-resistance

William Froude

graphs for each set of models were plotted for comparison. There was indeed a degree of similarity in the curves for each hull form across the scales, and each plotting demonstrates that above a certain speed the *Raven* or 'sharp' form showed greater resistance than the *Swan* or 'blunt' form. This was a significant finding since the *Raven* possessed a hull shape which was thought to guarantee a resistance lower than any other shape. The plotted results seemed to demonstrate that this was wrong, a claim that Froude was anxious to test.

The River Dart experiments with the *Swan* and *Raven* were conducted sporadically until early 1868 when Froude communicated the results to the Admiralty. He was able to demonstrate that the

resistances between models varied as the cube of the scaling factor, but was unable to give a convincing reason as to how the speed related to the resistances. This dilemma was solved during the summer and autumn of 1868 when he developed what later became known as Froude's hypothesis. This stated that two components combined to make up ship resistance, the first was water friction on the surface of the hull and the second the differences in pressure over the hull that resulted in the generation of the wave system that accompanied the hull. Since the generation and maintenance of these waves took energy from the ship's engines they represented a form of resistance called 'wave resistance'. Using this idea and his knowledge of water friction Froude showed that the two resistance components did not scale in the same way. He further demonstrated that the similarity condition for the wave resistance across scales was met, providing the waves made by each scale of model diverged from the model's bows at the same angle.

The arguments that Froude presented to the Admiralty to justify his ideas were more complex than the above explanation might suggest but they were of sufficient authority to be accepted as sound. On this basis the Admiralty constructed the world's first ship model testing tank in 1870 at Chelston Cross, Torquay, adjacent to Froude's house. The tank was so successful that it was soon followed by others constructed by both private ship builders and foreign governments. From the time of its inauguration the Chelston Cross tank was used routinely to examine the resistance properties of all new hull designs proposed by the Admiralty and in addition was employed in fundamental research into the design of propellers and a host of other hydrodynamical problems relating to the design of ships and the prediction of their performance. TW

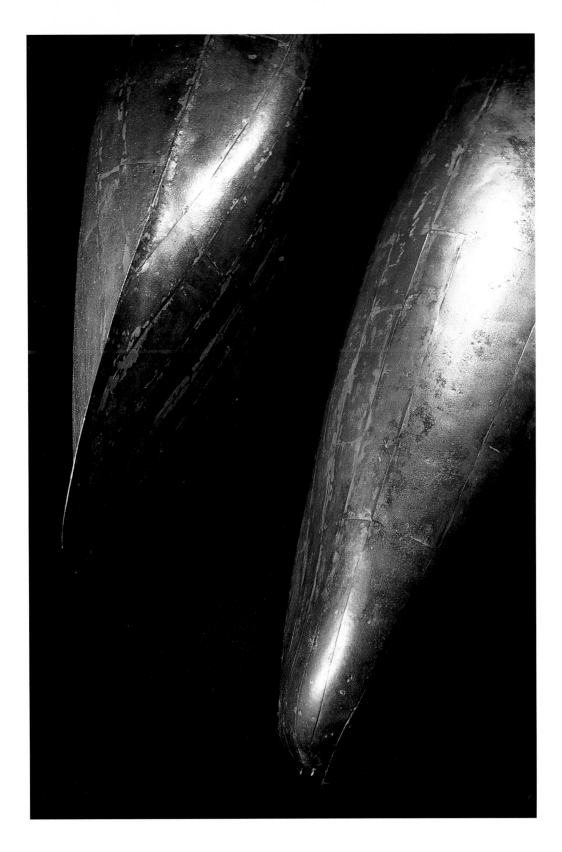

HOLMES' LIGHTHOUSE GENERATOR

One of the greatest breakthroughs in nineteenth-century science was Michael Faraday's fundamental discovery that an electrical current is produced when a wire cuts through a magnetic field. He discovered this effect, known as magneto-electric induction, in August 1831. In November of that year this celebrated British chemist used this principle to construct a disc generator capable of producing a relatively continuous electric current. In so doing, Faraday opened up a completely new field for the development of mechanical devices that transformed energy continuously from a mechanical to an electrical form.

A stream of magneto-electric generators emerged in quite rapid succession over the next thirty years, each commonly known after its maker and each usually boasting an improvement in efficiency or output over its predecessor.

In about 1850, F Nollet (1794–1853), Professor of Physics at the Military School of Brussels, was working in lighthouse illumination. He intended to produce a brilliant light (limelight) by heating a lump of lime in an oxy-hydrogen flame. To produce the necessary oxygen and hydrogen gases, he built a crude disc armature machine with coils and magnets intended to break down water by electrolysis. The experiment was not a success. In 1853 Nollet died and Frederick Hale Holmes (fl. 1840–75), an analytical chemist and engineer from London, was invited to assist. Holmes' background is obscure, but he re-sponded and became associated with the Compagnie De l'Alliance, the company set up to develop Nollet's work.

In Paris while working on these magneto-electric generators, Holmes realized that they might provide a continuous current capable of producing a brilliant electric arc which could be used for lighting. By 1855 he had built a number of successful machines, and acquired sufficient knowledge, experience and confidence to solicit the Elder Brethren of Trinity House, the English lighthouse authority, to test his magneto-electric generators and arc lamps for lighthouse illumination. Trinity House sought Michael Faraday's advice and agreed to a trial at Blackwall in 1857 under his direction. The generator was five feet long, five feet wide, four and a half feet high and weighed two tons.

The experiment proved successful and with Faraday's backing, two larger Holmes machines were installed in the South Foreland high lighthouse. It was here on the Kent coast, on 8 December 1858 that electric light first became fully operational in a lighthouse. Faraday's role as consultant for Trinity House was both influential and instrumental in a further Holmes machine being taken into use at Dungeness in June 1862. This machine would run (intermittently) for the next thirteen years.

The Holmes magneto-electric machine in the Science Museum was designed in 1867, and was one of a pair put into service on 12 January 1871. This was at Souter Point, a new lighthouse erected the year before on the east coast between South Shields and Sunderland. Prior to installation this machine was exhibited at the Paris Universal Exhibition of 1867, and afterwards erected at Trinity Wharf, Blackwall for experimental purposes.

Michael Faraday's wholehearted endorsement of Holmes' machines directly influenced Trinity House's receptiveness to the electric light, and the adoption of Holmes' system in 1857. But by the 1860s it was no longer clear that Holmes was ahead in the development of magneto-electric machines. Ironically, it was the appearance of Holmes and this machine at the Paris Universal Exhibition in 1867 that did much to bring to the attention of the English lighthouse authorities the significant progress made by the French in this field.

From a purely technological standpoint, it was quite remarkable how this dinosaur of a machine ever reached the installation stage at Souter Point as late as 1871. Perhaps Holmes' long association with Trinity House, the latter's innate conservatism, and a sense of loyalty to Faraday (who had died in 1867) combined to see Souter Point lighthouse lit by Holmes' 1867 machine. This one remained in service there until 1900, but long before then the market was dominated by dynamo makers Siemens of Germany, Gramme of France, and Brush of America, the latter providing a range of complete arc lighting systems for quite different applications.   DJW

JULIA MARGARET CAMERON'S 'IAGO'

The world's earliest photographs, taken over a century and a half ago, needed very long exposures. None were portraits, for nobody could keep sufficiently motionless for the minutes, even hours, required. But, as soon as the two prototype systems – the English calotype and the French daguerreotype – reduced exposure times to less than sixty seconds, things changed: perhaps as many as ninety-five per cent of all the daguerreotypes ever taken were portraits (*See* TALBOT'S LATTICED WINDOW, and EARLY DAGUERREOTYPES OF ITALY).

During the 1840s, scores of portrait studios were opened, first in London and Paris, then in cities large and small throughout the world. Their owners, more interested in quick profits than aesthetics, rushed customers through. Not surprisingly, most of the resulting pictures have little claim to any artistic merit. There were, of course, exceptions. In Scotland, the likenesses of hundreds of scientists, artists, clerics and leaders of Edinburgh society were captured by the painter David Octavius Hill and his partner Robert Adamson, in carefully posed and lit studies. Taken in the mid-1840s, these were perhaps the first masterpieces in the history of photography.

Twenty years later, a plain, bossy, ambitious and middle-aged English gentlewoman was given a camera by her daughter and son-in-law, in the hope that it would keep her busy while her husband visited his coffee plantations in Sri Lanka, then called Ceylon. Their hopes were more than realized – Julia Margaret Cameron (1815–79) became the greatest photographic portraitist of her time.

Mrs Cameron used Frederick Scott Archer's 'wet collodion' process (*See* LEWIS CARROLL'S PHOTOGRAPHS). The operation was complex, difficult, even dangerous, especially for a woman who had been looked after by servants all her life. Not surprisingly, Cameron was hugely proud of her first portrait. The note she sent to the father of the young girl sitting so still in it (for perhaps a minute or two at a time) gives

Sir John Herschel by Julia Margaret Cameron

some idea of the scale of the problems, and of her triumph at solving them:

Given to her father by me . . . My first perfect success in the complete Photograph . . . This Photograph was taken by me at 1 p.m. Friday Jan. 29th. Printed – Toned – fixed and framed all by me & given as it now is by 8 p.m. the same day.

A photograph needing such a long time, and so much effort, was unlikely to be as stereotyped and dull as a commercial portrait. The struggle to master the chemistry, and to keep the sitter still throughout a lengthy exposure, helped many nineteenth-century amateur portraitists produce far more memorable results than their professional counterparts.

Cameron lived on the Isle of Wight, a neighbour of Alfred Lord Tennyson, Queen Victoria's Poet Laureate. Through him, she met and photographed many notables of the day, among them Browning, Carlyle, Darwin and Herschel. To the latter, who introduced her to the finer points of photographic technique, and whom she called her 'Teacher and High Priest', she gave one of the finest albums of her pictures. Among her portraits of Tennyson and other distinguished Victorians, 'the peasantry of our island' and her own family, is the unique *Iago, Study from an Italian*. No other example is known and it may be that, after making one print, she accidentally broke the negative; it often happened.

Iago seems to be the only example of Cameron using a professional model. Alessandro Colorossi was one of a group of London models, mostly men, who satisfied Victorian artists' demands for Italians and their 'feeling for classical stance'. One of the painters who employed him was Cameron's friend George Frederic Watts and in an early version of Watts' *The Prodigal Son*, Colorossi can be seen with the bristly beginnings of a beard. The resemblance to *Iago* is unmistakable.

Perhaps Watts asked Cameron to take the photograph for his reference. Or perhaps she visited him while Colorossi was posing in his studio. She was certainly fond of Italian subjects, and there are two others in the Herschel album.

Mrs Cameron's life-size close-ups have a powerful directness unequalled in portraiture. Compared with the tiny *carte de visite* portraits then in vogue, they are almost overwhelming in scale and personality. When she photographed great artists and scientists, she was consciously 'recording faithfully the greatness of the inner as well as the features of the outer man', and she captured the personality of her Italian model just as effectively. Was she aware of something evil as he stood absolutely still before her bulky camera? Or did she sense it as the murky image slowly emerged in the darkroom? Only then, perhaps, did she decide to call this compelling picture *Iago*. CJF

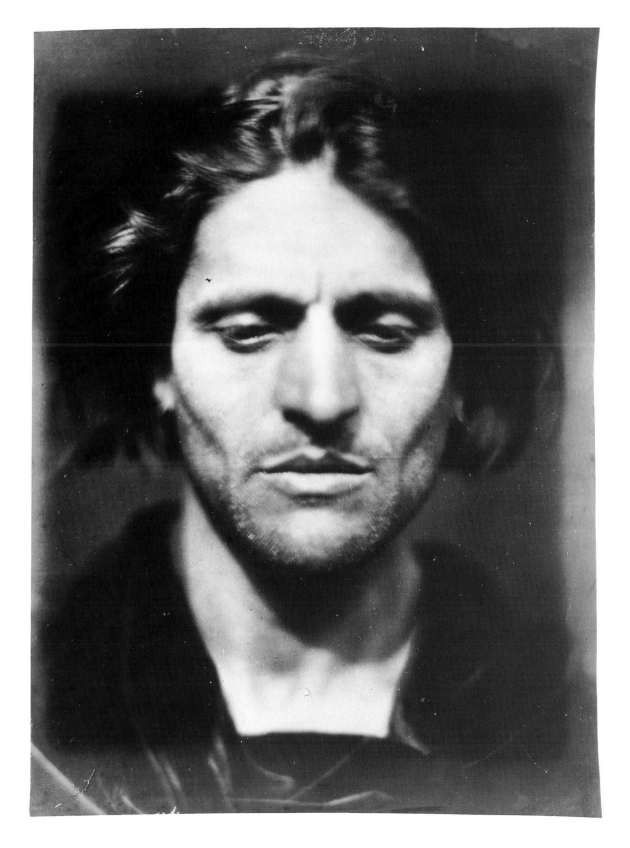

CROOKES' RADIOMETER

Cathode ray tubes are widely used today, they form the screens in television sets, computer terminals, radar sets and measuring instruments.

The characteristic glow produced by passing an electric current through air at low pressure has been known since the early eighteenth century. In 1855 Heinrich Geissler (1815–79), a glassblower in Bonn, devised a new pump which achieved a much higher vacuum. Geissler made the colourful Geissler tubes which delighted lecture audiences, but he also made apparatus used by Julius Plücker (1801–68) at the University of Bonn. Plücker noticed that at very low gas pressures the electrical glow discharge disappeared, to be replaced by a green fluorescence from the glass at one end of the discharge tube. This fluorescence was caused by rays emanating from the negative electrode, or cathode.

William Crookes (1832–1919), an experimental chemist and physicist, came to the subject by a different route. In 1861 he discovered a new element, which he called thallium. Over the next ten years he measured its atomic weight very carefully. Many samples had to be weighed in a vacuum, and Crookes noticed certain inconsistencies in the weighings: warm bodies seemed slightly lighter than cooler ones. He set out to examine the effect, partly out of scientific curiosity and partly because he supposed it might be connected with psychic and paranormal phenomena, which fascinated him.

This work led Crookes to devise, in 1875, the 'light-mill' or radiometer. It consists of a vane with four arms, each blackened on one side, balanced on a point in the centre of a glass bulb from which most of the air has been removed. When light or radiant heat falls on the vane it rotates, the blackened faces moving away from the radiation. This effect depends on the blackened faces of the vane absorbing more radiation than the other faces and so becoming slightly warmer.

In some experiments Crookes passed an electric current through a radiometer, which produced the familiar glow from the residual gas in the bulb. When

William Crookes, cartoon by Spy from *Vanity Fair*, 1903
Opposite: Maltese Cross tube and Radiometer

the vane was made negative, there was a dark space between it and the glow. The dark space was larger on the blackened faces of the vane. Believing that it could help his understanding of the radiometer, Crookes began a new series of researches on this 'Crookes dark space'. As the gas pressure was reduced, the dark space increased until it filled the whole of the bulb, and the fluorescence seen by Plücker appeared. Crookes investigated the cathode rays which caused this fluorescence.

Crookes reported this work to the Royal Society in the Bakerian Lecture of 1879. Subsequently he was asked to lecture at the Royal Institution and to the British Association for the Advancement of Science.

It was the climax of his career. His demonstrations aroused enormous interest. Cathode rays caused minerals such as ruby to phosphoresce, as well as the glass of the discharge tube. The stream of cathode rays could be deflected by a magnet, showing that they were electrically charged, but otherwise they travelled in straight lines. This was shown using an obstacle in the shape of a Maltese Cross, to cast a shadow in the fluorescence at the end of the discharge tube. Over a century later a form of Maltese Cross tube is still used to demonstrate the properties of cathode rays, which are now known to be beams of sub-atomic particles called electrons.

The radiometer has never been more than a popular scientific toy but Crookes' work on cathode rays was seminal. In 1895 Wilhelm Conrad Röntgen (1845–1923) noticed that when cathode rays struck the end of a discharge tube, rays of a new kind were emitted, capable of penetrating matter. He called them X-rays, and their significance can hardly be overstated. Crookes was disappointed not to have discovered X-rays himself.

In 1897 John Joseph Thomson made measurements which suggested to him that cathode rays might be streams of particles each carrying a single unit of charge but much lighter in weight than any atom. This is regarded as the discovery of the electron, and the whole of modern electronics depends upon it (See THOMSON'S CATHODE RAY TUBE).

Also in 1897, Ferdinand Braun (1850–1918) used a form of Crookes tube as a measuring instrument. It was the direct antecedent of the modern cathode ray tube, though it needed the discovery of the thermionic emission of electrons from a heated filament before it could take its final form.

From about 1881 Crookes' former assistant Charles Gimingham used his skills in glassblowing and vacuum technique to help Joseph Swan put his electric filament lamp into production, ushering in the age of electric lighting. The electric lamp in turn led to advances in electronics. Thus the ramifications of Crookes' work are enormous.   CNB

BELL'S OSBORNE TELEPHONE

In 1981 the American communications corporation AT&T came up with the slogan 'Reach out and touch someone'. The idea was simply to get customers to make more use of the telephone, but the phrase evoked echoes of an incident that had occurred in England 103 years earlier.

Alexander Graham Bell (1847–1922), inventor of the telephone and effectively the founder of AT&T, came from a Scottish family obsessed with speech and hearing. His grandfather, father and uncle all ran speech and elocution practices. His father married a deaf girl who became Bell's mother. Bell in turn became a speech practitioner in Boston, Massachusetts and married one of his clients, a young woman who was profoundly deaf.

It was possibly this familiarity with the problems of the deaf which led Bell into his royal *faux pas* of 1878. Bell had been invited to demonstrate the telephone to Queen Victoria at her residence, Osborne House on the Isle of Wight. On the evening of 14 January, contact was duly made with nearby Osborne Cottage and also with Cowes, Southampton and London. Among the telephones used was the instrument now in the Science Museum.

At some critical point of the demonstration, the Queen turned away from her telephone. Bell, used to attracting the attention of the deaf, leaned over and touched her arm, becoming the first commoner in years to lay hands on royalty without permission. Queen Victoria may or may not have been amused, but it certainly gave Bell unlooked-for publicity in court circles. 'Reach out and touch someone,' the slogan might have read, 'preferably royalty.'

The walnut and ivory decoration lavished on this crude telephone by instrument makers Julius Sax and Company indicates how eager Bell was to impress. By 1878 the telephone, revealed in America in 1876, was familiar to British technical specialists through the lectures and demonstrations given by the British physicist Lord Kelvin (1824–1907) and Sir William Preece (1834–1913), Engineer-in-Chief of the Post Office. But the big push towards commercial success was just starting, and Bell needed all the patronage he could get. He employed one of the world's first public relations advisers, the American journalist Kate Field who, as well as being on hand to sing down the telephone to Queen Victoria from Osborne Cottage, kept newspapers on both sides of the Atlantic amply supplied with material.

Field's accounts picture the Osborne demonstration as a jolly entertainment, with singing and laughter heard distinctly over the wires, but it was in reality only a partial success. The line from Cowes failed, and the line from London was so bad that Bell did not attempt to transmit speech, confining himself to the tones of an organ. There were clearly many practical problems still to be overcome.

One of these was the question of the transmitter. The Bell telephone used identical devices for transmitter and receiver, each consisting of an iron diaphragm, a magnet and a coil of wire. Sound waves made the diaphragm move, producing electric currents in the coil and sending electric waves down the line to the receiver. Here the effect was reversed, with currents in the coil moving the iron diaphragm, recreating the original sound. However the system lacked any amplification, making its range frustratingly limited. It was Thomas Edison (1847–1931) who, a month after the Osborne demonstration, patented the amplifying carbon microphone and provided a partial solution to the problem of transmitting telephone signals over long distances.

This instrument is one of several thought to have survived from the Osborne demonstration. There are three in the Smithsonian Institution in Washington, and one at the AT&T Archive in Warren, New Jersey, although all these lack the handsome wooden switchplate provided by Sax. This one is labelled 'Osborne Cottage' and was probably used not by Queen Victoria but by her equerry Sir Thomas Biddulph and his wife. They are reported as having conversed with the Queen, while Kate Field, after her vocal contribution, had to make do with Queen Victoria's thanks relayed by the Duke of Connaught. Following the demonstration, the Queen, in spite of describing the telephone in her journal as 'rather faint', was sufficiently impressed to ask Bell if he would sell two of the instruments left at Osborne. This appears never to have happened.

We take the telephone so much for granted now that it is difficult to recapture the *frisson* that Bell's tinny simulations of human speech must have elicited. Still less can we imagine the social confusion created by this unfamiliar visitor who could enter a house without being presented at the front door. Or perhaps we can. How did you feel when you first saw or used a mobile 'phone? And how will you feel when such devices are common in shirt pocket or handbag? The coming era of truly personal telecommunication, as well as resting on Bell's fundamental invention, will enable us to appreciate its impact on those who, like Queen Victoria at Osborne, witnessed it when it was new. RB

This drawing, from a circular produced by Bell's agent Reynolds in 1877, shows how in his original system the same design of instrument was used as both transmitter and receiver

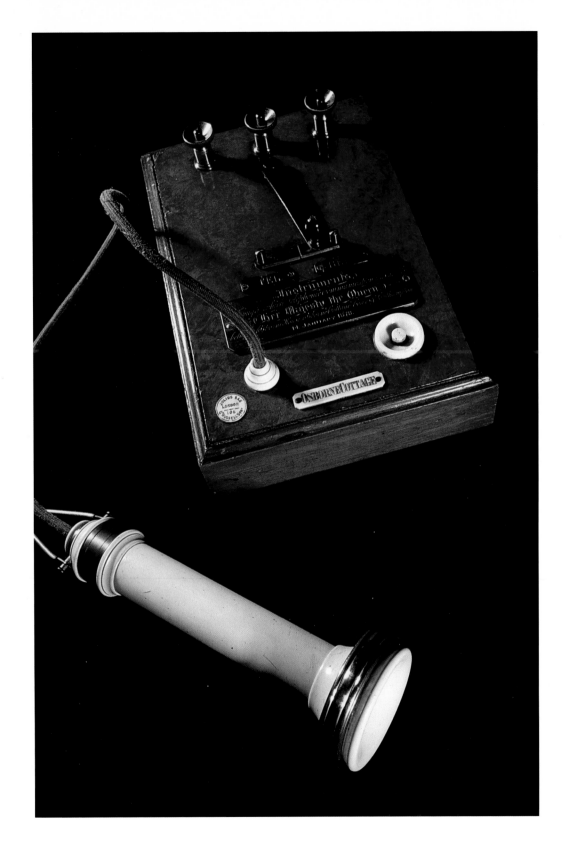

WIMSHURST'S ELECTROSTATIC MACHINE

Victorian culture saw science as an endless source of wonder and inspiration. Social soirées took the form of demonstration evenings where the middle classes would learn of the latest ingenious devices and tricks. Ordinary people were interested to learn as much as they could about scientific developments. Electricity was a particular source of fascination. If a thing could be done then electricity could do it.

Static electricity, caused by a build up of electrically charged particles, was the first form of electricity known to man; the word electricity derives from the Greek for amber, a substance which readily develops an electrostatic charge when rubbed.

In Victorian England amber was replaced by glass or sealing wax which, when rubbed with silk, could demonstrate the basic principles of static electricity by attracting small pieces of paper or feathers. Many Victorian devices or toys such as the electric puppet, the electric cannon and wheel were used to demonstrate the mysterious qualities of this form of electricity.

In 1860 the need for reliable high-voltage machines for applications in telegraphy and laboratory experiments stimulated experiment.

Friction, as a method of producing high-voltage static electricity, was unreliable. The electrophorus, invented by Volta in 1775, used another method called induction. The development of induction machines in Continental Europe and Britain culminated in a machine designed by James Wimshurst (1832–1903) in 1882. Known simply as the Wimshurst machine it completely revolutionized the science of static electricity. It was the first machine capable of producing high-voltage static electricity that was not affected by atmospheric humidity.

When the relative humidity is high, electrostatic phenomena, such as the electric charge created

The 'electric bath'
from the *Journal of Electro-Therapeutics*, May 1893

when you rub a balloon against a woollen sweater or walk across a synthetic rug, nearly disappears. This is because layers of moisture on the surfaces allow electrical charges to 'drain off' to earth and not accumulate. Likewise many electrostatic machines could not accumulate a charge and would not produce electricity in humid conditions. Wimshurst's machine was the first to work in almost all atmospheric conditions. It featured other improvements that made it highly reliable and easy to use.

Wimshurst attributed this remarkable success to the fact that he had not followed what he called the 'orthodox views of electrical science' and worked with complete originality of thought. His machine

has not been improved on significantly since.

The two fourteen-and-a-half-inch-diameter discs rotated in opposite directions, with positive charge building up on one and negative charge on the other. When discharged they produced sparks that could jump a distance of four and a half inches. Wimshurst's largest machine had discs of seven feet in diameter and produced sparks that could jump a distance of two and a half feet, corresponding to several hundred thousand volts.

At the turn of the century electrostatic machines such as the Wimshurst were used to generate high voltages to power X-ray tubes. Another common application was in electrotherapy. A patient would be charged up in an electric bath or subjected to electric shocks. These treatments, although considered very efficacious for insomnia, rheumatism, sprains, bruises and even constipation, were soon replaced by less dangerous and more effective methods. Most conditions once treated by electrotherapy are now treated by chemical means.

Nowadays electrostatics has become important because of its uses and its dangers. Electrostatic precipitators are used in industry to remove contaminating particles from gases, and in domestic air conditioners to remove dust. Photocopiers and laser printers employ the principles of electrostatics as do spray painting and crop spraying devices. However precautions have to be taken to prevent build up of electric charge on aircraft, electronic equipment and in situations involving inflammable substances such as filling petrol tanks.

James Wimshurst was of the amateur scientist breed, although he was a marine surveyor by profession. His obituary states that what he did 'he did for the love of work and advancement of science – all for love and nothing for reward.'   PJB

THE FIRST TURBO-GENERATOR

Look here, you fellows, I have an engine that is going to run twenty times faster than any engine today.

The Hon C A Parsons

Charles Algernon Parsons (1854–1931) was the youngest of three sons of the Earl of Rosse, an accomplished Victorian astronomer (*See* THE ROSSE MIRROR). The family seat at Birr Castle was a focus for scientific thought and discussion among those who came to use the telescope. Astronomers employed by the Earl to operate his observatory by night were additionally required to tutor his children by day. Charles and his brothers enjoyed regular access to the estate workshops and grew up in an atmosphere of lively intellectual enquiry allied to practical engineering skills. These were nurtured in Charles' case by two years at Trinity College, Dublin followed by a four-year degree course in mathematics at Cambridge. Much of his time as an undergraduate was devoted to mathematical experiments. He was particularly preoccupied by the limitations on speed and size which then confronted the builders of reciprocating steam engines as they sought new ways of exploiting the increase in boiler pressure then becoming available. Remarkably for a young man of his privileged background, C A Parsons enrolled in 1877 as a premium apprentice at Armstrong's works at Elswick, where he perfected his mechanical skills.

In 1881, he obtained an appointment with Kitson's, engineers of Leeds. Throughout this period his persistent intellect was wrestling with the engineering problems of steam turbine design. The idea was far from novel; James Watt's steam wheel was one of many hundreds of attempts to harness the expansive power of steam to produce rotative power direct, without the intervention of a piston which wasted energy. It was known that to obtain efficiency, the vanes of a steam turbine should travel at about half the speed of the steam passing over them. Steam escaping to atmosphere at a pressure of 200 pounds per square inch can achieve a velocity of 4,000 feet per second. Parsons realized he could

A portable Parsons turbo-generator, 1885.
A similar generator was used to light open-air ice skating in January 1886, as a novel way of raising money for a Newcastle hospital (NEI Parsons)

harness such high velocity if he divided the pressure drop into several stages so that the reduction in the steam's velocity was gradual and work could be extracted from its expansion at each stage.

Parsons settled on fifteen progressive stages of expansion for his first practical steam turbine, which he completed in 1884. Sets of fixed vanes attached to the turbine casing separated the rings of moving vanes attached to the turbine shaft, directing the steam on to them and helping to accelerate the flow of steam. The machine ran at 18,000 revolutions per minute (rpm), unprecedented for a steam engine. Parsons recognized the need to limit the diameter of the rotor so that centrifugal forces remained manageable and the engine remained balanced. A neat solution to the problem of end-pressure on the axial turbine was achieved by admitting the steam at the mid-point of the turbine shaft and dividing the flow into two equal streams, the axial loadings from which balanced each other, so that no thrust bearing was necessary. The turbine shaft thus had to be slender to minimize the external diameter of, and hence the centrifugal force on, the rotors and it had to be long to accommodate a total of thirty sets of rotors and guide rings, since there were fifteen stages of expansion on either side of the steam inlet.

Such a configuration posed problems of transverse vibration as the shaft was accelerated through a series of critical speed ranges corresponding to the natural aural frequencies of the shaft when struck. Once through these critical speed ranges, the 'whirling' phenomenon abated and acceleration could safely continue. Parsons saw that the whirling would be exacerbated if the shaft bearings sought to restrain the shaft altogether, but could be dampened if the bearings permitted a degree of controlled deflection through the critical speed ranges.

Elegance in engineering is often characterized by simplicity and Parsons' answer to the whirling problem exemplifies this. In his first 1884 turbine the bearing journals at each end of the rotor shaft enjoy about $\frac{1}{32}$ inch of latitude. Each is supported by a large number of washers alternately fitting the bush and the casing. A spring compresses each assembly so that relative movement between adjacent washers is inhibited. Each bearing can align itself to the direction taken by the revolving shaft, but sudden and destructive lateral displacement of the shaft is controlled by the friction between the washers.

Parsons ran his turbine for extended periods at 18,000 rpm and proved the adequacy of the balancing and lubrication arrangements. To accommodate any whirling of the shaft, clearances between the rotor and the turbine casing were generous, and steam consumption was consequently prodigious. But because it demonstrated that sustained high-speed rotation was safely obtainable from a steam engine, the Parsons turbine enabled generators to be greatly simplified and the rotating armature made smaller now that it could be rotated so much faster than previously.

The turbo-generator patented by C A Parsons in 1884 represents the most important single development in the history of electric power generation. Today ninety-eight per cent of electricity consumed in the United Kingdom comes from steam turbines, using exactly the same principles as those demonstrated in 1884.

JCR

ROVER SAFETY BICYCLE

It is not always appreciated that the design of the present-day bicycle has remained much the same as that originated in the 1880s. From the early 1870s to the late 1880s the high bicycle, Ordinary, or Penny-Farthing bicycle as it was commonly known, was extremely popular and led to the formation of numerous cycling clubs. By 1882 there were almost 350 clubs in the provinces and 184 in Greater London. Besides club members there were many more who took to the rough roads at weekends exploring countryside previously impossible to reach in a day.

However, the design of the Ordinary had many faults. It was unstable because the cyclist was almost directly over the centre of the large front wheel and there was the danger of the rider 'taking a header' if, for example, the front wheel hit a stone. It was difficult to mount and dismount, and the front wheel was both driven and steered at the same time which tired the rider's arms. The larger the diameter of the front wheel, the faster the speeds achieved because for every turn of the pedals a greater distance was covered. The only factor limiting the diameter of the front driving wheel was the length of the rider's legs.

The availability of better materials and the improved technology of components such as chains led to the search for a safer design of bicycle. When first introduced, the new design was so much lower and more stable than its predecessor, the high bicycle or Ordinary, that it became known as the safety bicycle.

There followed a number of experimental designs, some of which were produced commercially. Though the front wheel became smaller, it remained larger than the rear wheel and therefore still resembled the Ordinary. One of these was the Bicyclette patented by Henry Lawson in 1879 which had chain drive to the rear wheel.

In 1885 many rear-driven safety bicycles were shown in public for the first time. Amongst these was a Rover safety bicycle designed by John Kemp Starley

A contemporary advertisement for
the Rover safety bicycle, 1888

(1854–1901) and exhibited at the Stanley Show in London from 28 January to 3 February, 1885. J K Starley worked as a cycle mechanic until 1878 when he started his own business in Coventry with William Sutton. In November 1896 the Rover Cycle Company Limited was formed. From this evolved the Rover Company Limited, which produced many successful models of bicycles, motor cycles and cars. Cycle production ceased in 1926.

The Rover bicycle in the Science Museum is one of Starley's improved designs made in late 1885 and

may be regarded as the prototype which set the trend for future technical development and commercial production. The essential advantage of this design is the diamond-shaped frame which gives structural strength and stiffness combined with low weight and compactness. Smaller diameter wheels of almost equal size ensure a low riding position for far greater stability and a geared-up chain drive to the rear wheel allows more efficient pedalling. In 1885 a 100-mile road race was organized for riders of Rover bicycles, in which the existing 50- and 100-mile records were broken.

Although the Rover marked a turning point in the development of the bicycle there was still much more to be done before the bicycle in its classic form evolved. The steering is direct and the steering head slopes but the forks are straight, not curved. The frame was not yet triangulated as, for example, there is no tube from the saddle to the crank bracket. Pneumatic tyres were re-introduced in 1888 and were in general use a few years later, together with gears and better brakes. The Rover weighs thirty-seven pounds, compared to about twenty-eight pounds for modern touring bicycles and nineteen pounds for road racing bicycles.

It was, then, J K Starley's Rover safety bicycle that changed the course of cycle development, as by the early 1890s very few Ordinary cycles appeared in manufacturers' catalogues; it was then that the derogatory name of Penny-Farthing was first used. Most of the bicycles produced since the 1890s have been based upon the original Starley Rover design. As J K Starley said in a paper presented at the Society of Arts in 1898, '...my aim was not only to make a safety bicycle, but to produce a machine which should be the true *Evolution of the Cycle*, and the fact that so little change has been made in the essential positions, which were established by me in 1885, prove that I was not wrong in the cardinal points to be embodied to this end.'   FR

THE KODAK CAMERA

Just over one hundred years ago, there appeared a camera that was to change the course of photography. Popular photography can properly be said to have started in 1888 with the introduction of the Kodak.

The Kodak camera was the invention of an American, George Eastman (1854–1932). Advertised as 'the smallest, lightest and simplest of all Detective cameras' (a popular term of the 1880s for hand-held cameras), it was a simple wooden box six and a half inches long, three and three-quarter inches high and three and a quarter inches wide. It was small and light enough to be held in the hands while in use.

The name 'Kodak' is now synonymous with popular photography and is a household word all over the world. It was the success of the original Kodak camera which laid the foundations for Eastman to build an enormous international business empire. He chose the name for his new camera with great care: 'The letter "K" had been a favourite with me – it seems a strong, incisive sort of letter. It became a question of trying out a great number of combinations of letters that made words starting and ending with "K". The word "Kodak" is the result'. Eastman later explained to the British Patent Office: 'This is not a foreign name or word; it was constructed by me to serve a definite purpose. It has the following merits as a trade-mark word: first it is short; second, it is not capable of mispronunciation; third, it does not resemble anything in the art and cannot be associated with anything in the art.'

Taking a photograph with the Kodak camera was very easy, requiring only three simple actions: turning the key (to wind on the film); pulling the string (to set the shutter); and pressing the button (to release the shutter and make the exposure). It was, in many respects, the forerunner of today's point-and-shoot cameras. The lens fitted to the Kodak had a considerable depth of field, objects being in focus from as close as four feet, and giving a wide (sixty degree) angle of view. This meant that no viewfinder was needed, the camera was simply pointed at the subject to be photographed. Poor definition at the edge of the image area, however, meant that a circular mask had to be used in the camera, placed in front of the film. This accounts for the distinctive round (two-and-a-half-inch-diameter) photographs which the Kodak camera produced.

Ingenious, compact and simple to use though it was, the Kodak camera did not embody any revolutionary technological innovations. It was not the first hand camera, nor indeed was it the first camera to be made solely for roll film. The true significance of the camera, which makes it a landmark in the history of photography is that it was the first stage in a complete system of amateur photography. In Eastman's own words: 'The Kodak camera renders possible the Kodak system whereby the mere mechanical act of taking a picture, which anyone can perform, is divorced from all the chemical manipulations of

A snapshot taken with the Kodak camera c 1889.
The man on the left is holding a Kodak camera

preparing and finishing pictures, which only experts can perform … We furnish anybody, man, woman, or child, who has sufficient intelligence to point a box straight and press a button … with an instrument which altogether removes from the practice of photography the necessity for exceptional facilities, or in fact any special knowledge of the art.'

The Kodak camera was sold already loaded with enough film to take 100 photographs. After the film had been exposed, the entire camera was posted to the factory where it was unloaded and the film developed and printed. The camera, reloaded with fresh film, was then returned to its owner together with the negatives and a set of prints. Previously, photographers had no choice but to do their own developing and printing, needing a darkroom and the knowledge and skill to perform complex chemical manipulations. This, more than any other factor, had delayed the popularization of photography. With the Kodak system, Eastman had not only removed this barrier but also founded the modern developing and printing industry. To promote the Kodak camera, Eastman devised the brilliantly simple sales slogan that summed up his new system: 'You press the button, we do the rest.'

This new convenience, however, did not come cheap. In Britain the Kodak camera sold for five guineas (£5.25). The developing and printing service cost a further two guineas (£2.10). In 1888, £1 was a week's wage for many workers. Through a combination of pioneering mass-production methods and imaginative marketing techniques, Eastman was able to continue the successful process of turning photography into a truly popular pastime. Improved versions of the Kodak camera were produced and costs were greatly reduced. In 1900 the five shilling (25p) 'Brownie' camera was introduced. For the first time the pleasures of photography had been brought within reach of practically everyone.　CWH

EARLY CINE-CAMERAS

Louis Le Prince and the Lumière brothers have both been credited with the invention of cinema. Rarely should individuals be credited with absolute invention. The single lens camera built by Le Prince c 1888, and the Lumière Cinématographe developed in 1895 illustrate the point well.

Cinema can be described as the projection of moving photographic pictures to an audience. Its evolution was dependent on a handful of technical principles. In 1832 Joseph Antoine Ferdinand Plateau (1801–83) constructed a device which created the illusion of movement through the successive presentation of still images showing phases of that movement. Photography, the permanent record of optically-formed images on light-sensitive material, was perfected simultaneously by Louis Daguerre (1787–1851) and William Henry Fox Talbot (1800–77) in 1839 (See TALBOT'S LATTICED WINDOW and EARLY DAGUERREOTYPES OF ITALY). Thus, the technical principles for cinematography were essentially understood by that date.

By the 1870s, a number of people were experimenting with the recording and analysis of movement using photographic techniques. Photographic emulsions allowing exposures as short as one thousandth of a second were available, but only on glass plates. By the mid-1880s the key issues were the need for long strips of pictures and a method of moving the strip intermittently at a fast enough rate to record movement smoothly – around sixteen pictures per second. In 1885, George Eastman (1854–1932) introduced a paper-based roll film (See THE KODAK CAMERA).

The Le Prince single lens camera made use of Eastman's paper roll film to record a sequence of images. It is claimed that in October 1888, it was used to take twenty consecutive pictures of Leeds Bridge at a rate of about sixteen pictures per second. The camera and the frames still exist, although we have no definite proof of their date. But an English patent applied for by Le Prince (1842–1890?), a French showman engineer and inventor, on 10 January 1888 describes the principles of cinematography.

October 1888 was also the month French physiologist Etienne Jules Marey (1830–1904) gave a presentation to the Académie des Sciences in Paris. He showed a series of pictures made at a rate of twenty a second on a roll of Eastman paper film. The chronology is further complicated in that on 17 October 1888, Thomas Alva Edison (1847–1931) filed a caveat with the American Patent Office describing a picture version of the phonograph.

Le Prince disappeared in mysterious circumstances in 1890. One theory suggests that Edison had him murdered, although recently discovered papers suggest that Le Prince was worried by large debts – a less dramatic but more plausible explanation. It is possible that this single lens camera was the first mechanical expression of cinematography, but it may also have been faked by Le Prince's son Adolphe who, in 1898, testified against Edison's claims to be the inventor of moving pictures.

It was, however, indisputably Edison who introduced moving pictures to the general public when the first Kinetoscope parlour opened in New York in 1894. The Kinetoscope was a coin-operated machine which gave a 'show' lasting about twenty seconds for a single viewer – the original peepshow.

The brothers Louis (1864–1948) and Auguste (1862–1954) Lumière were the most successful photographic plate manufacturers in France. They first saw a Kinetoscope in the summer of 1894. Impressed by the demonstration but put off by the high prices demanded by Edison's agents, they decided to develop their own product. In February 1895, they patented a combined camera and projector which used an intermittent claw derived from the mechanism used in sewing machines (See ELIAS HOWE'S SEWING MACHINE) to move the cloth.

The apparatus was called the Cinématographe and the first public presentation was made at the Société d'Encouragement pour l'Industrie Nationale in Paris on 22 March 1895. It showed a one minute film of workers leaving the Lumière factory in Lyons. Encouraged by its reception, further films were made and for the first time on 28 December 1895 an audience paid to see projected, moving photographic pictures in the basement of the Grande Café, Paris.

The Lumières commissioned an initial production run of 200 Cinématographes from the instrument maker Jules Carpentier. This Cinématographe is No.8 from that run and is the earliest known example. It was purchased in Paris in 1896 and was immediately exported to Paraguay where it stayed in the same family until being sold in 1990. RV

Above: Lumière Cinématographe No 8 of 1896.
Opposite: Le Prince single lens camera c 1888
(the top lens is a viewfinder)

LOCOMOTIVE FROM FIRST TUBE RAILWAY

When it opened in December 1890, the City & South London Railway was the world's first deep-level underground railway and the first true electric railway in that it was a line with defined stations and a proper system of signalling. It represented an enormous advance both in tunnelling and in electrical engineering but paradoxically was not designed with electricity in mind. The design of the original locomotives has perhaps as much historical importance to electric railway traction as Stephenson's *Rocket* had to the development of steam railways sixty years earlier.

In the mid-nineteenth century cities throughout the world were experiencing traffic problems little different from those of today, except in scale. In 1863 London pioneered the urban underground railway as an answer to congestion. Engineering constraints meant that the tunnels were constructed by 'cut and cover' – digging a deep trench which was then lined with bricks, covered over and the top surface restored. This involved extensive demolition of buildings and disruption of traffic and essential services; despite this, many miles of subterranean railway were built in this way in London. Their ability to ease congestion was only partial, not least because of the necessity to use steam locomotives, creating a severe ventilation problem.

In the 1860s an eminent engineer, P W Barlow (1809–85), assisted by his pupil, W H Greathead (1844–96), developed a shield for tunnelling through the London clay at a sufficient depth to avoid existing services and building foundations, while causing minimal surface disturbance. With the opening of a short tunnel under the Thames in 1870, the Tower Subway, deep-level underground railways became a practical possibility. Despite many proposals for such railways, by the mid-1880s only one scheme had succeeded in getting beyond the planning stage. This was the City of London & Southwark Subway, which was planned to haul the trains using a continuously-moving cable driven by a stationary steam engine. With Greathead as engineer, construction of the

tunnels began in 1886; in 1887 an extension southwards to Stockwell was agreed, bringing the route mileage to about three and a half miles.

However, the failure early in 1888 of the company supplying the haulage cable equipment forced a re-examination of other possible means of traction. Electricity was still a comparative novelty, and in London there was only a small number of power stations, each supplying electric light to a few local consumers. Some lightweight electric railways existed but nothing on a heavy scale had yet been tried. It was thus a courageous decision of the CLSS to opt for electric traction. It was at first intended to place the motors under the carriages themselves, but in January 1889 the company accepted a proposal from the Salford firm of electrical engineers, Mather & Platt, to use locomotives instead. They were also to supply the other electrical equipment for the line, including the power station. Two prototype locomotives were built, followed by twelve others.

The first prototype was designed by Mather & Platt's chief engineer Dr Edward Hopkinson (1859–

City & South London Railway locomotives at Stockwell Depot, 1890. The first prototype, No 1, with a shorter cab, is on the right

1922), in conjunction with his brother Dr John Hopkinson (1849–98), who was retained by the firm as a consultant. It had a shorter cab than the others. In December 1890 the City & South London Railway, as it had recently been renamed, started services using these fourteen locomotives together with thirty carriages built by the Ashbury Carriage & Iron Co Ltd. For the first year or so the locomotives were run in the livery in which they had been delivered, an unlined reddish-brown. The C&SLR then devised a standard colour scheme of black with yellow/orange panels and bright yellow lining-out, and the locomotives were dealt with as repainting became due.

After an unsteady start, traffic built up and the line was extended in both directions. (It now forms part of London Underground's Northern Line.) This required more locomotives which embodied improvements in the electrical equipment and could haul longer and heavier trains but which were otherwise similar in looks and layout. In addition, Numbers 3 to 12 of the first batch were rebuilt between 1904 and 1907 with more powerful motors. Numbers 1 and 2 were withdrawn around 1901 and Numbers 13 and 14 followed in about 1907. The withdrawn engines were stored at the depot at Stockwell, being cannibalized for spares as required.

The locomotive now in the Science Museum was retained when the C&SLR was rebuilt from 1922 to 1924 and the original trains scrapped. However, its identity had been forgotten: it bore the maker's plates from Number 1 and was painted in reddish-brown. It was displayed in this guise for over sixty years until it was finally decided to restore the more representative livery and the correct identity, which surviving records suggested was almost certainly Number 13. The work was carried out in 1990.

Locomotives of essentially this design lasted in service for over thirty years; a remarkable record, given that there had been nothing quite like them before. Number 13, in its authentic bright colours, perhaps only now properly commemorates the pioneering achievement of the C&SLR's creators. JL

PARSONS MARINE STEAM TURBINE

By the start of the last decade of the nineteenth century the multi-cylindered steam expansion engine had become the major power plant used to propel ships. The principle of expanding steam and hence driving a piston to do useful work, by means of passing it through three or more cylinders of ever increasing diameter and lower pressure had reached the stage where the physical limitations of the process were apparent. In a period that saw Britain's proportion of world merchant ship steam tonnage climb to fifty per cent, the ever increasing demands for more efficiency and more power to drive bigger and faster naval and merchant ships created an important market for improved types of marine propulsion.

One candidate for this market was the steam turbine which relied for its operation on steam passing through two sets of blades, or winglets, one set being fixed to the static engine casing and the other to a disc attached to the shaft of the machine. The effect of passing steam through this arrangement was to make the disc and hence the shaft rotate. By repeating these stages the steam was allowed to expand continuously; this led to a much faster, lighter and smaller engine in comparison with the expansion engine.

In January 1894 the Marine Steam Turbine Company was formed in Newcastle upon Tyne to exploit the developments of the Hon C A Parsons. In 1884 he had constructed a steam turbine to provide the motive power to drive a dynamo generating electricity (See THE FIRST TURBO-GENERATOR). In 1889 the partnership which had built the 1884 machine was dissolved. Under the terms of the partnership agreement, the patents relevant to the many improvements Parsons had introduced during the period 1884–9 became the property of his former partners. This meant that in particular he was not allowed to design a turbine that incorporated parallel flow, that is, one that allowed steam to flow through the engine in the same direction as its shaft or axis.

To overcome this Parsons resorted to using the idea of radial flow whereby the steam was introduced near the shaft and then made to flow radially outwards before being conducted back inwards towards the shaft where it could be used for another outward pass. It was this type of steam turbine which was constructed by the Marine Steam Turbine Company in 1889 and which was used to propel the world's first steam turbine driven vessel, *Turbinia*.

To estimate the power needed to drive *Turbinia*, Parsons conducted a series of resistance experiments with two models of the ship in early 1894. The predictions of the experiments were that the turbine would have to develop 820 horsepower at a boiler pressure of approximately 200 pounds per square inch to drive the vessel at thirty knots, using a single propeller shaft rotating at 1,600 revolutions per minute. Preliminary speed trials of the *Turbinia* were conducted from 14 November 1894 but with disappointing results: a speed of just under twenty knots was achieved, ten knots below specification.

The *Turbinia* at speed; Parsons is the tall figure just forward of the funnel

Over a further series of experiments using a specially developed torsion dynamometer which measured the power being developed by the turbine and transmitted by the propeller shaft, the fault was diagnosed to lie with the design of the propeller. Although seven different designs of propeller were tried over thirty-one trials no improvement in overall performance was achieved.

After numerous experiments the problem was found to lie in the fact that the drop in pressure over the back of the propeller blades made the air dissolved in the water form cavities. These cavities led to a drop in efficiency and a loss of thrust; this phenomenon was later termed cavitation. Ultimately the cavitation was caused by the inability of the propeller to absorb the large power being generated by the turbine due to its small diameter and small surface area.

In order to overcome cavitation it was decided to replace the single turbine, single propeller shaft arrangement with three turbines each driving their own propeller shaft. This decision was made towards the end of 1895 and it coincided with the recovery of Parsons' original patents. The radial turbine was then taken out, three parallel flow reaction turbines installed and newly designed propellers fitted. It was essentially this arrangement that drove the *Turbinia* at 32.75 knots in February 1896 and which so impressed the Navy when it appeared at the Naval Review later on in the year.

Although the radial turbine was never used again after its trials on the *Turbinia* and its design was superseded by the parallel or axial type, this particular engine was the seminal engine that not only demonstrated the effectiveness of the steam turbine for marine propulsion but also the need to understand the phenomenon of cavitation in the design of propellers.   TW

PANHARD ET LEVASSOR CAR

On 7 March 1910, the Science Museum acquired its first motor car, presented by the Royal Automobile Club. In recommending its acquisition the then Curator of Land Transport, E A Forward, wrote, '... had we to choose only one car from among the many historical cars existing, then a Panhard car such as this would have the best claim.' Indeed, the Panhard was considered to be as significant to motoring as the *Rocket* locomotive had been to railways.

In Britain, Lanchester, Wolseley and others were experimenting with cars in the mid-1890s. However, the successful invention, development and manufacture of motor cars was a continental phenomenon. In Germany Benz had demonstrated the first practical car in January 1886. The most successful petrol engines were the high speed engines built by Daimler, also in Germany, from 1884, and these were used in motor cars from 1886. But it was the French Panhard et Levassor (using the Daimler engine) that transformed motoring into a serious proposition.

Panhard et Levassor made woodworking machinery in Paris and had acquired the French rights to the Daimler engine in 1889. They first experimented with cars in which the engine was in the centre or at the rear of the vehicle. Then in 1891 they offered for sale a car in which a front-mounted, vertical, two-cylinder engine drove through a flywheel, clutch and three-speed gearbox to the rear axle. The driver and passengers sat between the engine and the rear axle. This arrangement became the standard layout for the motor car for the ensuing seventy years.

A car of this type attracted world-wide publicity when it was the first to finish in the first road race from Paris to Bordeaux and back on 11–13 June 1895. In a great feat of skill and endurance the car was driven single-handed by Emile Levassor himself. He covered the distance of 732 miles in two days forty-eight minutes, at an average speed of 15 mph and arrived six hours before the next competitor. Panhard cars were to dominate the great Continental road races until 1900.

Meanwhile in Britain, the embryonic motor industry was restricted by legislation which required motor vehicles to be preceded by a pedestrian to warn of their approach, effectively limiting their speed to 4 mph. The Hon Evelyn Ellis (1843–1913) was one of a small number of pioneer British motorists who had the vision to see the importance of the motor car and campaigned to have the law amended. Ellis was a Member of Parliament and the son of the sixth Baron Howard de Walden. He chaired the inaugural meeting of the Automobile Club of Great Britain and Ireland (from 1907 the Royal Automobile Club) and became its first Vice-Chairman. He was also a Director of the Daimler Motor Co Ltd of Coventry.

On 25 June 1895 Ellis bought a Panhard of the type which Levassor had driven in the Paris–Bordeaux–Paris race. It had engine number 394 Type P2D of three and three-quarter horsepower with coachwork by Belvalette. It was the first car imported into this country and its first journey from Micheldever sta-

The Hon Evelyn Ellis in the driving seat of his Panhard, at the first public exhibition of cars in Great Britain; Tunbridge Wells, 1895 (Veteran Car Club of Great Britain)

tion to Ellis' home at Datchet on 5 July was vividly described by F R Simms in the *Saturday Review*:

We set forth at exactly 9.26 a.m. and made good progress on the well-made old London coaching road ... We were not quite without anxiety as to how the horses we might meet would behave towards their new rivals, but they took it very well, and out of 133 horses we passed on the road only two little ponies did not seem to appreciate the innovation ... It was a very pleasing sensation to go along the delightful roads towards Virginia Water at speeds varying from three to twenty miles per hour ... There we took our luncheon, and also fed our engine with a little oil ... Going down the steep hill leading to Windsor, we passed through Datchet, and arriving right in front of the entrance hall of Mr Ellis's house at Datchet at 5.40, thus completing our most enjoyable journey of fifty-six miles, the first ever made by a petroleum carriage in this country, in 5 hours 32 minutes, exclusive of stoppages.

Ellis claimed to have driven the Panhard for more than 2,000 miles over the next eighteen months but was never prosecuted for exceeding the speed limit and not being preceded by a pedestrian. It was one of the five cars at the first public exhibition of cars in this country at Tunbridge Wells on 15 October 1895. It also took part in the Emancipation Run from London to Brighton on 14 November 1896 which was held to celebrate the coming into force of the Locomotives on Highways Act which freed motor cars of the need to be preceded by a pedestrian and raised the speed limit to 12 mph. From this moment motoring in Britain was able to develop.

Ellis sold the car for about £3 in 1908. It was acquired by Mr A Vaughan-Williams, a consulting engineer, who exhibited it in a display of historic cars in the Imperial International Exhibition at White City in London in 1909. Recognizing its importance, the RAC bought the car for £50 and presented it to the Science Museum the following year.   PRM

'COMING SOUTH, PERTH STATION'

Coming South, Perth Station is one of a pair of paintings by the Victorian artist George Earl (1824–1908). This painting and its companion *Going North, King's Cross Station* capture the spirit of the railway age at the height of its prosperity, before the advent of the motor car, when everyone who travelled, travelled by train. Apart from William Powell Frith (1819–1909), in his seminal work *The Railway Station* (Paddington) 1862, and George Earl in these two works, no other Victorian painter was to represent in such panoramic detail the world of railway station interiors: their architecture, the trains, the wide range of passengers and their imagined stories, contemporary advertisements and station staff.

George Earl is primarily known as an animal and sporting painter, who exhibited at the Royal Academy from 1856 to 1883 and at the British Institution and the Society of British Artists. Although Earl was perhaps best known for his dog portraits, often in simple formats, he also attempted elaborate and sometimes dramatic compositions in which figures, animals and landscapes are combined.

Earl originally explored the railway themes in two earlier works exhibited at the Royal Academy in 1876 and 1877. Later, reputedly at the request of Sir Andrew Barclay Walker of the Walker Brewery, he painted the two enlarged and amended versions of his original compositions. Both paintings came into the possession of the Walker Brewery Company and until 1990 hung in one of the Walker public houses, The Vines, in Lime Street, Liverpool.

This painting with its companion, *Going North, King's Cross Station*, are fine illustrations of the social impact of the railway at the zenith of its power and influence in the late Victorian period. The two paintings fit neatly between the impressionistic works of the twentieth century and the more romantic and sentimental canvasses of the nineteenth century, both of which are already represented in the National Railway Museum's picture collection.

In the painting *Coming South*, dated 1895, the scene is Perth station. The fine train shed designed by Sir William Tite in 1848 and lit by elegant gas lights is a prominent feature. Although the station at Perth was enlarged in 1885–7 Earl shows it in its earlier condition with up and down trains leaving from the one hall. During conservation work on the painting, an examination of the brush strokes confirmed that Earl was responsible for depicting the architectural features of the station as well as the dogs and crowd scenes. This is interesting given that Frith is known to have had assistance from William Scott Morton with the architectural setting of *The Railway Station*.

A train bound for London waits to depart. The first carriage is marked for London, Euston. A group of passengers, with their hunting dogs, game, bags and other equipment prepare to leave Perth following a successful shooting houseparty. Among the luggage lie grouse, blackcock and some stags' antlers whilst a porter heads towards the train with a rabbit and a wicker basket doubtless containing salmon. Station porters are again in evidence struggling with luggage on a trolley near the second carriage, the centre portion of which would be the baggage compartment. A boy walks the station selling *The Scotsman* newspaper and the grooms and footmen control the dogs as best they can. On the left an elderly woman in red, standing behind a man in a tam-o'-shanter hat appears upset at the departure of their daughter.

The station clock shows 15.50 which indicates that the train waiting to depart was the 16.04 arriving in London at 03.50 in the morning. The train would follow the West Coast route from Perth to London. The Highland Railway terminus is shown in the centre of the painting. The Highland Railway train would leave at 16.30 for Inverness, arriving at 22.05.

Among the several gun dogs Earl shows in the central group are pointers, which flank two English setters and a pair of Irish setters with gleaming tawny coats. Interestingly, there is a well behaved border collie profiled on the right looking on, behind the cloaked woman: this is not a gun dog and accompanies the gentleman in the top hat who is perhaps present to witness the departure of a friend. The platform is littered with feathers, food and an open book possibly dropped by the woman in peach who is bending as though to retrieve it.

An interesting feature of the Perth painting is the prominence of the railway station signs. A notice at the end of the Highland Railway bay points people to the Caledonian Railway for through trains to Aberdeen. The left luggage rooms and the ladies first class waiting rooms are clearly indicated. A little further down the platform the ladies second class waiting room sign can just be seen. On the pillars alongside the platform are instructions forbidding smoking on the station. This is believed to have been an early instruction to passengers for safety reasons which was soon relaxed.   CJH

Purchased with the assistance of the National Heritage Memorial Fund,
the National Art Collections Fund, and private funds

THOMSON'S CATHODE RAY TUBE

Matter is made of atoms. The concept is so familiar that it is now part of general knowledge, yet a century ago some leading scientists doubted the very existence of atoms. The extraordinarily successful journey that has led to our present understanding of atomic structure began in the rather dingy laboratory of the first explorer-in-chief, John Joseph ('J J') Thomson (1856–1940). An enthusiastic and ambitious professor at Cambridge University's Cavendish Laboratory, Thomson used results he had taken with his cathode ray tube to prove in 1899 the existence of the first sub-atomic particle – the electron. In many ways, this discovery marked the birth of the modern electronic age.

Thomson was the apotheosis of the top-flight Cambridge scientist. He began his career by winning a scholarship to Trinity College and went on to be a Fellow, Professor and Master of the College, winning *en route* a Nobel Prize (1906), a knighthood (1908) and the Presidency of the Royal Society (1915). He was Professor of Experimental Physics, although that may not have been the best job for him; he was very mathematically minded and was well known to be an inexpert experimenter, to the extent that his students were fearful when he approached their equipment (his wife would not let him do even minor jobs about the house). He was nonetheless a brilliant designer of scientific apparatus and equally good at interpreting the results of experiments. Both of these strengths were displayed in his conclusive contribution to the cathode ray debate which was exercising European scientists in the 1890s.

In the late 1860s, it had been demonstrated that a column of gas at low pressure glows colourfully when electricity is passed through it from one metal plate to another. If the pressure was sufficiently low, the individual tracks of so-called 'cathode rays' could be seen between the plates (*See* CROOKES' RADIO-

J J Thomson with a cathode ray tube, 1897
(The Cavendish Laboratory, University of Cambridge)

METER). What did these rays consist of? If this question had been put to a European scientist in the mid-1890s, the answer would have depended to some extent on which side of the Rhine the scientist lived. German scientists held that the rays were mysterious waves in the ether which were supposed to pervade the whole of space, whereas French and British scientists believed that the rays were made up of individual particles.

In the race to rule out the wrong theory of cathode rays, Thomson made a key contribution in the spring of 1897. With the help of his assistant Everett, he constructed special cathode ray tubes that enabled him to study how the paths of the rays are affected by electric and magnetic fields.

In his Royal Institution lecture of 30 April 1897, Thomson boldly – some would say rashly – suggested that cathode rays were corpuscles even smaller than hydrogen atoms. At least one member of his audience thought he was joking! But he certainly was not and two years later he went further and suggested that cathode rays were indeed particles, each

with a definite charge and mass (his values for both turned out to be quite accurate). He had written his name into every book on atomic science as the discoverer of the electron.

This may have been a great event for the scientific cognoscenti, but would it make any difference to the lives of the public at large? Some scientists evidently hoped not; for some years after Thomson's breakthrough, his colleagues at the Cavendish toasted at their annual dinner to 'The Electron: may it never be of use to anybody'. These pompous hopes have not been fulfilled – electrons are now very much a part of modern life. They illuminate the screen of every television and personal computer and are responsible for the operation of every electrical device. In laboratories, electron microscopes allow us to map the shapes of viruses and the surfaces of atoms (both far too small for ordinary optical microscopes); the electronics industry aims to exploit new and more ingenious ways of controlling electron flow; and radiologists routinely use electrons from radioactive decays to diagnose and treat cancers.

Yet while the aspirations of the Cavendish purists have been thwarted by the technologists, others have advanced our knowledge of the sub-atomic world beyond Thomson's dreams. The electron is now known to be but one of several fundamental particles, each envisaged to exist at an unimaginably small point, with neither shape nor size. Scientists now study these basic building blocks of matter using huge particle accelerators that are each, in a sense, descendants of Thomson's little cathode ray tube. Today's particle physicists normally work in multinational teams, taking years to design and carry out experiments that consume alarmingly large chunks of national research budgets. The days in which a lone scientist could make a bench-top discovery of a new particle seem, sadly, to be long gone.   GPF

MARCONI'S FIRST TUNED TRANSMITTER

It has been said that Mr Marconi has done nothing new. He has not discovered any new rays; his transmitter is comparatively old; his receiver is based on Branly's coherer. Columbus did not invent the egg, but he showed how to make it stand on its end . . .

William Preece, Friday Evening Discourse
at the Royal Institution, June 1897

In 1896 Annie Marconi, well heeled daughter of the Jameson Irish whiskey family, brought her son Guglielmo (1874–1937) to England to seek a sponsor for his new system of wireless telegraphy, the Italian Ministry of Posts and Telegraphs having shown little interest in it. His first success was with the prickly and conservative Engineer-in-Chief of the Post Office, William Preece, the man who had once dismissed Bell's telephone (*See* BELL'S OSBORNE TELEPHONE) with the famous remark, 'I have one in my office, but more for show, as I do not use it because I do not want it.'

Preece became Marconi's champion, arranging and publicizing tests of wireless and even renouncing his own rival system. Marconi clearly had something that Bell, and several other pioneers to whom Preece offered less than wholehearted support, did not. Whatever it was, this relatively untutored youth progressed from experimenting in an attic in Bologna to producing this unglamorous but tuned, and in that sense modern, transmitter in just five years.

Heinrich Hertz (1857–94) had confirmed the existence of electromagnetic waves in 1888, generating them with the help of sparks. A year later, Oliver Lodge (1851–1940) showed how an electric circuit could be made to vibrate like a plucked string and make a nearby circuit vibrate in sympathy. In 1894, when he was twenty, Marconi started experimenting with Hertz's waves with the express intention of using them as tools of communication.

Five years on, with wireless threatening to degenerate into Babel because of interference, Marconi was exploiting Lodge's ideas to produce a solution. The qualities that attracted Preece to Marconi, perhaps because he shared them, were evident

throughout. Weak on theory, but with a clear vision of his goal, Marconi groped his way to the realization of what academics had merely shown to be possible. Prototypes like the one in the Science Museum were his indispensable guides.

The tuned transmitter emerged from Marconi's work on detecting rather than creating electromagnetic waves. From the beginning he had used as part of his receiver the coherer, a well known but capricious device which he turned into a reliable component. But at the end connected to the receiver, his aerials produced large currents and small voltages; the coherer needed the opposite.

He found a solution in connecting the coherer through a special electrical transformer which his staff called the 'jigger'. It was the outcome of a programme of trial and error that began in 1897. By

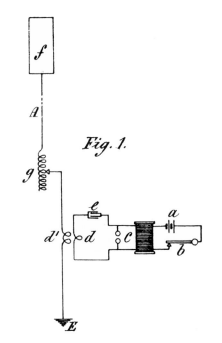

Fig. 1.

Marconi's 'jigger' (marked d'-d in this drawing from Patent 7777 of 1900) made it possible to transmit radio waves of well-defined frequency

1898 the jigger had increased by a factor of nine the distance over which Marconi could communicate, to sixty miles. Such sensitivity had its disadvantages. There could be many stations in a sixty-mile radius, and if more than two started up at once nobody would receive anything intelligible.

Someone in a crowd can concentrate on one speaker at a time, ignoring other conversations. The problem was to give a radio receiver similar competence. Marconi was not alone in realizing that with a transmitter producing waves of well-defined frequency, the electrical resonance that Lodge had demonstrated could make a receiver respond strongly to that transmitter and no other. The difficulty was finding a means of producing electromagnetic waves at a desired frequency. Solving this problem was the final step towards tuning and modern radio communication. The jigger transformer, originally designed to improve reception, provided the means.

Like Hertz, Marconi used an electrical spark as a means of releasing radio waves. With the spark gap connected directly to the aerial, as in the early Marconi transmitters, radio waves were produced in bursts too brief for any definite frequency to be discernible. By connecting the gap through a transformer – the few turns of wire wrapped round the frame of this prototype – each spark could produce a train of waves long enough to excite resonance. If all the circuits involved were tuned to what Lodge called 'syntony', transmitter and receiver could be alone in a crowd.

William Preece, wrong about so many things, was right about Marconi, the gifted magpie. In fulfilling his vision, Marconi used whatever and whoever came to hand, but in the way that an artist must use his materials, without prejudice. He did not stumble upon the tuned transmitter, but he did not arrive at it quite rationally either. It was first revealed in his Patent No. 7777 of 1900. Numerologists may make of this what they will, but without doubt this humble lash-up is a seminal fragment of the engineer's art. RB

POULSEN'S TELEGRAPHONE

Those who regard the telephone answering machine as the quintessential curse of modern existence will be surprised to learn that it was invented by a relatively obscure Dane as long ago as 1898. In the process, Valdemar Poulsen (1869–1942) also invented magnetic recording, the basis of a huge industry encompassing tape, discs and indeed answering machines themselves. He did not conceive the idea, nor perfect it, but he was undoubtedly the first with a working device.

When Poulsen was born in Copenhagen, few had ever heard speech from the mouth of a machine. By the time he died, magnetic recording, while not yet commonplace, was growing to maturity in the service of war. But the late nineteenth century was still too early to attempt such an invention. Poulsen's only option was to translate every subtlety of the sound wave into a corresponding fluctuation of the magnetic record, and to do this properly he needed amplifiers of a kind not then available. Digital recording, the idea that sound waves could be reduced to numbers, unsubtle and easy to record, was not seriously put forward until just five years before his death.

At the time Poulsen revealed the machine he was eventually to call the telegraphone, the culture of the industrialized world was still dominated by the written word. Edison at first thought of his own recording machine, the phonograph, as a device that would allow telephone calls to be handled by a central office, as telegram messages were handled by a post office. Poulsen's attitude was more democratic. In his first British patent, he describes machines that would sit at home waiting for their absent owner, announcing the time of his return to callers and if required taking messages.

Between 1898 and 1915 Poulsen produced many modifications of his original idea. His first machine was clearly a magnetic version of the phonograph, using a cylinder wrapped with steel wire which became magnetized in a pattern similar to the sound

Valdemar Poulsen never took a degree but was awarded two doctorates for his pioneering work in speech communication
(International Telecommunications Union)

waves. He later used reels of wire for recording incoming calls, and also considered steel tape, discs and even magnetically coated paper. But he struggled with the problems of handling these media.

So as well as lacking amplification, his machines were complicated and probably unreliable. They achieved some success however for office dictation. Indeed their quality of reproduction gained Poulsen the Grand Prix of the Paris Exposition of 1900, though it must be said that the only contemporary standards of comparison were the horribly scratchy phonograph and gramophone.

Although no records of the manufacture or sale of the Science Museum's telegraphone have survived, we know that it was used in one of the Royal Dockyards, and its design indicates a date of 1903 or 1904. The machine used steel wire driven at about fifty times the speed of modern tape. It could be used as either an answering or a dictating machine. It would have been supplied with a pair of Bell-type telephones, and both had to be pressed to the ears to hear speech at intelligible volume. But users were impressed by the lack of background noise, making possible the recording of whispers and sighs, something the rival phonograph could not match.

Magnetic recording did not really become useful until three fundamental problems had been solved. Poulsen never really tackled any of them. To start with, amplification is needed, and plenty of it. The power from the playback head in a typical cassette system is increased 100 billion times on its way to the speakers. Then there has to be a medium that will retain fine magnetic detail, so that the recorded wavelength of sounds can be reduced and with it the absurdly high wire consumption of the telegraphone.

And finally, to get an undistorted recording, high-frequency bias is required, to compensate for the fact that magnetic media do not react in the same way to both large and small signals. Poulsen failed to solve this problem completely, and a solution was not found until 1927. Paradoxically, the very best modern recordings, because they are digital, can be made without this.

Ultimately, the telegraphone was a failure; it was too much in advance of the technology of the day. A really satisfactory answering machine was not available until the 1950s (like the telegraphone, it recorded on wire) and it was thirty more years before 'Please speak after the tone . . .' became an everyday request. Ever the democrat, Poulsen gave up recording and instead developed the first radio transmitter capable of handling speech. This pre-electronic device enjoyed only brief acclaim. But the technology which replaced it went on to turn his other little failure into a huge success.   RB

FLEMING'S ORIGINAL THERMIONIC VALVES

Distress and disappointment hardly seem an appropriate reward for the holder of one of the most significant patents in the history of electronics. Yet those were the misfortunes that befell John Ambrose Fleming (1849–1945) after registering his 'Improvements in Instruments for Detecting and Measuring Alternating Currents' in 1904. This title, and the ensuing unpleasantness, obscure the fact that Fleming's invention was the first practical thermionic device for detecting radio signals. Thermionic devices are those where the source of current is a heated negative electrode (the cathode), and the current flows through a vacuum to the positive electrode (the anode) via other electrodes which can be used to control the current.

Fleming's valve was the first of a line of devices which were to be the mainstay of electronics for its first half-century, and well into the 'solid-state' era, when transistors and integrated circuits inspired the miniaturization which we now expect. Even now, thermionic devices are not extinct. Cathode-ray tubes in television and computer screens (*See* CROOKES' RADIOMETER) and high-power valves in transmitters are examples; perversely, a few hi-fi purists eschew the predictable and consistent performance of solid-state amplifiers, preferring bulky ones with glowing valves.

John Ambrose Fleming became Professor of Electrical Engineering at University College, London, in 1885. He was knighted in 1929. Although an academic, he was the instigator of what might be called applied electronics. In 1882 Fleming had become 'electrician' to the Edison Electric Light Company and was soon absorbed with the task of improving carbon-filament light bulbs, which had a tendency to darken. Edison himself had already tried, unsuccessfully, to prolong bulb life by placing an extra electrode alongside the filament; he noticed that, with an extra electrode connected to the positive end of the filament, a small but measurable electric current flowed between them. This 'Edison effect' caused widespread curiosity and Edison patented his modified lamp as a novel 'electrical indicator'.

Fleming commenced more serious research into the Edison effect in 1889. He had experimental lamps made at the Edison and Swan (later Ediswan) Lamp Works. The results obtained with these, reported in 1890, led to further work in 1895 and 1896. By then, he had investigated electrical conduction in a vacuum between glowing filaments and a second electrode in a variety of shapes and sizes of tubes.

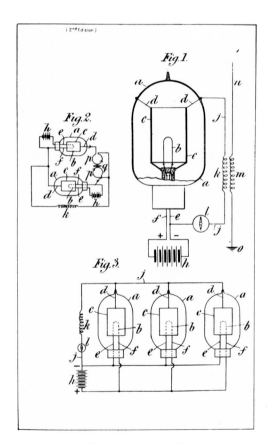

Part of the patent drawing of the first
wireless detector to use a thermionic valve.
Opposite: an experimental lamp which
Fleming used in 1889.
It was to one of these that he turned fifteen years later
for his wireless detector experiments

His painstaking account of this work is now regarded as a classic. But in a world which was more excited by Röntgen's revelations on X-rays and J J Thomson's identification of the electron (*See* THOMSON'S CATHODE RAY TUBE), Fleming's work was overshadowed.

But his moment was yet to come. Wireless telegraphy was expanding rapidly (*See* MARCONI'S FIRST TUNED TRANSMITTER). Fleming became a technical adviser to the Marconi Company in 1899, and helped design the transmitter at Poldhu, Cornwall, for the 1901 transatlantic transmissions. Receiving weak wireless signals, especially over such long distances, was a significant problem and Fleming, inspired in 1904 by 'a sudden very happy thought', turned to one of his earliest 1889 experimental valves for a solution. He linked it up to a simple circuit containing a battery, a meter, and suitable coupling to an aerial. He found the system worked well.

A patent was granted in 1905, but any benefits it might have brought Fleming became the property of the Marconi Company. 'Cat's whisker' detectors appeared a year later, and were in fact less cumbersome and equally efficient. And in 1907, the American Lee de Forest patented a detecting circuit using a valve with a third electrode. Fleming was well aware of the American's activities, and had fired warning shots to warn off the poacher in 1905. De Forest persisted, and a lifelong enmity between the pair of them ensued. However, it was not until 1912 that the full potential of the extra wire in de Forest's valve was realized, when it was incorporated into circuits which could generate and amplify speech and radio signals – a development which took communications firmly into the electronic era. But without Fleming's persistent work this transition would have been delayed – or credited elsewhere.

A final sad twist to the story was that enmity between the Fleming and de Forest factions in the United States spilled over into patent litigation, and in 1943 the US Supreme Court declared that Fleming's American patent was invalid. ED

HABER'S SYNTHETIC AMMONIA

Both science and quality engineering are important aspects of technological innovation. Nowhere is this lesson better demonstrated than in the story of the early-twentieth-century invention of synthetic ammonia, the basis of modern fertilizers. The two key relics of this German breakthrough are a tiny glass tube holding a few cubic centimetres of liquid and a thirteen-metre sixty-tonne steel pressure mantle. The sample in the tube was produced when the laboratory process was first demonstrated on 2 July 1909. The giant mantle is a relic of the industrial realization of the physical chemist's dream, and was part of a plant that operated from 1916 to 1982.

Ammonia had been compounded into fertilizers since the mid-nineteenth century, enriching the land to enable year-in, year-out food production for the world's growing population. But by the 1890s, it appeared that the main source, guano from Chile, would last no more than twenty years; the prospect was famine. Among the scientists replying to this challenge, the physical chemist Fritz Haber (1868–1934) calculated that it would be possible to obtain ammonia by simply 'burning' water, since the two ingredients nitrogen and hydrogen are found in air and water respectively. To convert his vision even into a laboratory demonstration required a new understanding of physical chemistry, high pressures, the assistance of catalysts made of the newly available osmium, and a well designed apparatus.

The apparatus was provided by a student-assistant of Haber from the Channel Islands, Robert le Rossignol. He designed a system which successfully produced synthetic ammonia. On 2 July 1909 the process was demonstrated to Alwin Mittasch of Badische Anilin- & Soda-Fabrik (better known today as BASF). Although only eighty cubic centimetres were produced that day, the principle had been proved and the company thereafter supported the commercial development of the process. A sample of that first demonstration production was kept by le Rossignol and later presented to the Science Museum.

Transforming an interesting laboratory demonstration into a commercial process destined to be the forerunner of all high-pressure chemical technology, was as formidable a task as the original discovery. However the company's belief that here lay a major commercial opportunity in agriculture was reinforced by military needs once war was declared in August 1914. Ammonia was the most practical source of the nitric acid needed to make TNT, the favoured general explosive, but a 1914 British naval victory, the Battle of the Falkland Islands, meant that Germany was cut off from supplies of Chilean guano.

A pressure mantle being installed in
the Oppau plant, 1920 (BASF)

As the scale of the requirement increased first to one tonne a day, then to twenty tonnes a day and then to 150 tonnes a day by 1916, the challenge of building the immense pressure reactors each capable of producing twenty tonnes of ammonia a day also grew. The combination of Haber's scientific achievement and the engineering project headed by Carl Bosch of BASF meant that by 1918 two full-scale plants at Oppau, near Ludwigshafen on the Rhein, and Leuna to the east, producing 50,000 tonnes each, were meeting fifty per cent of Germany's ammonia needs.

Wartime shortages meant that a new catalyst had to be found and an iron-based material was developed. It was also discovered that the hot hydrogen used to make ammonia reacted with the carbon in the steel of the reaction vessels at the enormous pressures needed, two hundred times atmospheric pressure. So a special lining of low-carbon steel had to be made, and to ensure this was not itself put under undue pressure the still cool but pressurized gases were circulated behind it.

In 1984 BASF gave the Science Museum one of the original mantles from the 1916 Oppau plant. It is made of two immense six and a half metre forgings bolted end on end. The thickness of the steel walls, twelve and a half centimetres, rises to thirty centimetres at the flanges. With the vessel, the Science Museum acquired the electric heater which brought temperatures up to the 200 degrees centigrade required, and a section of the holder with its powdered iron catalyst. Though these may appear much less glamorous than the intellectual achievement which won Haber the 1918 Nobel Prize for Chemistry, the equal importance of engineering was recognized in 1931 when Carl Bosch himself won the Prize for his contributions to the technology of high-pressure reactions. The process they developed is still in use today.   RFB

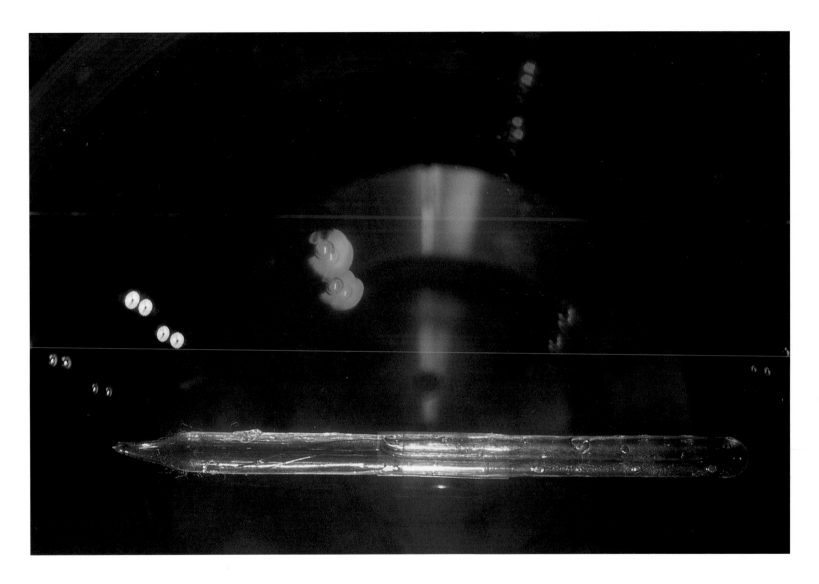

'THE MUNITION GIRLS'

'The people that will be able to keep its forges going, will perforce be the victor, for it alone will have arms', commented Messance on war in 1788. In August 1914 the likely scale and duration of world war were underestimated. Churchill's description 'A Steel War' in his Ministry of Munitions *Circular to Steelworks* of September 1917 captured the reality of the situation. During the Great War the iron and steel industry of South Yorkshire was turned over almost totally to the satisfaction of military needs.

The painting *The Munition Girls* by Stanhope Forbes, commissioned in 1918, depicts the manufacture of four-and-a-half-inch shells in the Kilnhurst Steel Works of John Baker & Co (Rotherham) Ltd. It is a record of an industrial process; red hot steel billets are reheated in an intermediate furnace and, still glowing, are moved by hand to a press for forging.

The gaunt and powerful press in the centre of the painting might seem ill-fitted for its purpose. However, the artist has recorded machinery originally designed for the production of forged tyres and axles for railway wagons and converted for shell making. In 1915 in common with many other steel makers, John Baker & Co had urgently to re-tool for armament manufacture and in particular to meet the overwhelming demand for shells for the Western Front. Such was the demand, that having refitted its own plant, John Baker & Co resorted to buying cattle cake presses and altering them for shell manufacture.

The dominant theme of the picture, however, is the workforce – women, the 'munition girls' of the title. The painting was commissioned by George Baker, Managing Director of the firm, primarily as '. . . a memento for our women workers, and each of them received a framed copy of it'. It is a graphic record of the wider employment of women in skilled trades. The call to the front had denuded the iron and steel industry of skilled male workers. Women left their traditional tasks, such as file cutting, and moved into the machine shops and foundries, operating equipment formerly only handled by men. Even the hot and heavy work of the foundry

was undertaken by women, though Forbes' painting appears to indicate that they were required in some numbers to equate to the muscle strength of their male counterparts. In the four years to November 1918 the number of women in the Sheffield steel industry increased sixfold. The 42,000 women in the nation's steel industry by the end of the war represented eleven per cent of the total workforce, some 36,000 more than before the war and a vital contribution to alleviating the labour shortage.

Both munitions manufacture and women's work were subjects covered by official war art schemes from 1916, which came to include artists of industrial

Recognition at Burlington House;
The Munition Girls was exhibited at the Royal Academy Summer Exhibition in 1919. It gave rise to this cartoon in which a couple of society girls recognize as studio models the two figures whom Forbes had placed in the right foreground of the painting

activity such as C R W Nevinson and Muirhead Bone, although women's work efforts received little official artistic attention during the war compared to those of men. In 1918 the first official British woman war artist, Victoria Monkhouse, was commissioned to record women in their new working roles, but to depict them clean and picturesque. Iron and steel companies in Sheffield and Rotherham independently commissioned artists to record their own war efforts, building upon a long-standing tradition of artistic patronage of foundry and manufacturing scenes. A rolling mill, and women making files, are among scenes by E F Skinner from a set of paintings commissioned in 1917 by Cammell Laird in aid of the Red Cross. However, Skinner's *For King and Country* of 1919, in the Imperial War Museum, is primarily an apotheosis of the heroic woman shell worker rather than an acknowledgement of women's contribution to the industry.

Stanhope Alexander Forbes (1857–1947) was not an official war artist. After study in London, work in Paris and visits in Brittany, in 1884 he settled in Newlyn, Cornwall, founding the influential Newlyn School of Art in 1899. An exhibitor at the Royal Academy in London from 1878, he was elected an Academician in 1910. *The Munition Girls*, exhibited at the Academy in summer 1919 contrasts markedly in subject and tonality with the landscapes, coastal and town scenes and figure studies for which Forbes is primarily known, but it retains the narrative content and social realism that were his hallmarks.

Forbes painted several industrial subjects, among which were *Forging the Anchor* (exhibited 1892), and paintings of the Sandberg sorbitic steel process. These works may have stimulated George Baker to commission *The Munition Girls* in 1918. Writing to Mrs Baker in 1940, Forbes recalled: 'It was indeed an unforgettable sight to see those fine women carrying on their work so splendidly, and the opportunity which Mr Baker gave me to record this wonderful service is one for which I can never be sufficiently grateful'.

WS

ASTON'S MASS SPECTROGRAPH

Francis William Aston (1877–1945) was presented with the Nobel Prize for Chemistry on 10 December 1922 for his work in establishing the isotopes of the non-radioactive elements. Isotopes are atoms which have the same chemical properties, but different atomic mass. For example, carbon can exist in its common form C^{12} or the less common form C^{14}. At the award ceremony it was acknowledged that while radioactive elements had readily shown the existence of isotopes, to prove that non-radioactive elements also possessed isotopes had been far from easy.

J J Thomson (*See* THOMSON'S CATHODE RAY TUBE), himself a Nobel Laureate, had achieved inconclusive experimental results to this end, but with the quartz microbalance and mass spectrograph, Thomson's assistant Aston identified two isotopes of the element neon and went on to employ his original mass spectrograph to find the different isotopes of about fifty elements.

Aston was the second son of a Birmingham metal merchant and farmer and in 1893 began to study science under Tilden, Frankland and Poynting at what later became Birmingham University. Whilst he continued to contribute to basic understanding in this area of research for the ensuing twenty years, it was his appointment by Sir J J Thomson at the Cavendish Laboratory in Cambridge that led to the discoveries by which Aston is now best remembered.

Aston was assigned to improving Thomson's apparatus in which a beam of positively charged particles (positive rays) were deflected by a combination of electric and magnetic fields into sharp visible curves, each representing an individual particle's charge-to-mass ratio. He thought that this apparatus gave rigorous proof that all the individual molecules of any given substance had the same mass. This Daltonian belief was rudely shattered in 1912 when Thomson obtained two curves for neon corresponding to masses 20 and 22. Two explanations suggested themselves: if neon had a true atomic weight of 20 (instead of the previously agreed figure of 20.20) then either mass 22 was an unknown compound of neon or a new element, meta-neon.

Aston was assigned to investigate the latter possibility and tried to separate the meta-neon by a variety of techniques. To see how well he was succeeding in separating neon and the mysterious substance, he devised the miniature quartz microbalance. The arm of the balance was level only when the case surrounding the balance was filled with gas of known density. Aston then filled it with the unknown gas, and altered its pressure, thereby varying its density until the tiny beam balanced again. Comparing the

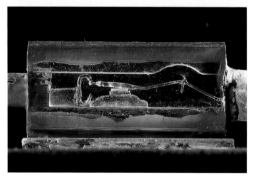

Top: Aston with his third mass spectrograph
(The Cavendish Laboratory, University of Cambridge).
Above: Aston's Quartz Microbalance

pressures allowed him to calculate the atomic weight of the unknown gas. The results indicated that the mysterious substance was an element with the same properties as neon, but with a different atomic weight.

The intervention of the war delayed further experiments but on returning to the Cavendish in 1919, Aston attacked the problem from a different direction, building a positive ray or mass spectrograph. The key advance over Thomson's apparatus was Aston's arrangement of the electric and magnetic deflecting fields so as to bring rays of uniform charge-to-mass ratio to sharp focus on a photographic plate. Aston devised several methods for calibrating his instrument and in the case of neon obtained mass lines on his photographic plate at 20 and 22 with the intensities of the lines showing that the two particles occurred in the ratio of 10:1, consistent with an average mass of 20.20, the known atomic weight of neon. Neon was thus proven to be isotopic and in the short time before Aston was presented with his Nobel Prize he had demonstrated the existence of isotopes in some thirty other gaseous elements.

Aston's work provided important insights into the structure of the atom and the way different elements are related to each other, and for these reasons he received the ultimate accolade of the scientific community.

With later more accurate mass spectrographs Aston obtained valuable information on the stability and abundance of particular isotopes. Through this work on the structure of the atom he early foresaw the dangers should the incredible energy of the nucleus be uncontrollably released, and lived long enough to see his forebodings given awesome substance. Meanwhile his principal experimental tool, the mass spectrograph, has been refined almost beyond recognition, adapted and adopted as an analytical instrument of prime importance in innumerable areas of chemical and biological research and industrial practice.   DAR

ALCOCK AND BROWN'S VICKERS VIMY

The Vimy aircraft which made the first non-stop flight across the Atlantic was presented to the Science Museum by Vickers Ltd and Rolls-Royce Ltd in December 1919, and has been the central feature of the National Aeronautical Collection ever since.

In 1913, the *Daily Mail* offered a prize of £10,000 to the first aviator to cross the North Atlantic. The outbreak of war in 1914 prevented any attempts to win the prize, but the offer was revived in November 1918. The Vickers company decided to enter the contest in February 1919, using a twin-engined Vimy aircraft flown by Captain John Alcock (1892–1919) and Lieutenant Arthur Whitten Brown (1886–1948), as pilot and navigator respectively.

The Vimy was designed as a heavy bomber to attack targets in Germany, in response to German bomber raids on London in 1917. Although the prototype Vimy was flown in November 1917, a shortage of engines prevented deliveries of production aircraft to the Royal Air Force until February 1919, after the Armistice.

The Vimy was a conventional design for its time, built mainly of wood with fabric covering, but careful attention to detail design and weight-saving produced an efficient structure capable of carrying the large fuel load needed for an estimated flight time of some twenty hours. The two Rolls-Royce Eagle VIII engines were probably the most reliable in service at the time; most contemporary types needed complete overhaul after about thirty hours running, but the Eagle's standard figure was one hundred hours.

This was perhaps the most important technical factor in the Vimy's successful Atlantic crossing.

The aircraft selected for the flight was modified during construction at Vickers' Brooklands factory, mainly to increase the fuel capacity from 516 gallons to 865, giving a nominal range of 2,440 miles. All military equipment was deleted, and the pilot's cockpit altered to accommodate the pilot and navigator side-by-side on a narrow flat wooden bench, softened by a thin cushion.

After a single test flight at Brooklands, the Vimy was dismantled and shipped to St John's in Newfoundland, where it arrived on 16 May 1919. The assembly of the aircraft in the open by a team of ten workers from Vickers and a single Rolls-Royce engineer, using improvised scaffolding and lifting gear, was no small feat. Finding a suitable field to fly from was also a problem; eventually, by demolishing a stone wall and removing several trees, a sloping run of 500 yards was contrived.

The aircraft was ready to leave on 14 June. Alcock and Brown arrived at the field just before dawn, but had to wait ten hours for the wind to drop. They were finally airborne at 1.41 pm (16.12 Greenwich Mean Time), climbing slowly to the west before turning back on to their course for Ireland and crossing the coast at 16.28 GMT. For eight hours, the flight proceeded between cloud layers above and below, but eventually Brown was able to get sextant sights on two stars and fix their position, well ahead of his estimate, revealing a tail wind of 40 mph.

Three hours later, Alcock lost control while flying through dense cloud at about 4,000 feet and the Vimy fell into a spiral dive, emerging from the cloud in a steep bank. Luckily, he was able to recover control just above the sea and resume course. At about 05.30 GMT, heavy snow obscured the fuel-flow sight gauge, and Brown had to climb up from his seat and kneel on the fuselage to clear the glass and confirm that the fuel pumps were still operating; he had to repeat this excursion from the relative warmth of the cockpit into the blizzard several times.

At 07.20 GMT, Brown obtained a sun sight fixing their position near the Irish coast. At 08.25 GMT they crossed the Irish coast near the Marconi radio station at Clifden, about twenty-five miles north of their intended landfall. After sixteen hours in the air, they were keen to finish the trip, and Alcock decided to land in a large green field not far from the radio station. Unfortunately, the field was actually a bog; the Vimy touched down at 08.40 GMT and came to an undignified halt with its nose in the mud.

Alcock and Brown returned to an enormously enthusiastic public welcome, and were knighted by King George V. The Vimy was dismantled and returned to Brooklands, where it was rebuilt using new components from store to replace broken parts. It was not flown again. Over the next few years, Vimys made several record-breaking long-distance flights, including England to Australia and England to South Africa. The design also formed the basis for a commercial airliner. JAB

AUSTIN SEVEN PROTOTYPE CAR

Many attempts to produce a small, reliable and economical car had been made since 1910. But it was not until the appearance of the Austin Seven, in July 1922, that this type of car became practical. Sir Herbert Austin (1866–1941) was clear that the car should be aimed at the person who could only afford a motor cycle and sidecar but aspired to be a car owner. He said that his new car would 'knock the motor cycle and sidecar into a cocked hat and far surpass it in comfort and passenger-carrying capacity … I cannot imagine anyone riding a sidecar if he could afford a car'.

Adding the sidecar to the motor cycle in 1903 had brought family motoring within the reach of those who could not afford the much more expensive motor car. Cyclecars, based on motor-cycle engines and parts, had begun to appear around 1910. Their low price made these very simple three- and four-wheeled cars popular in spite of their crudeness. Many cyclecar makers closed during World War One or in the economic depression of 1920–1. The rest succumbed to competition from the Austin Seven which was a properly designed small car built in large enough numbers to offer real motoring at only slightly greater cost. Within ten years there were twice as many cars as motor cycles on the road and the cyclecar had almost disappeared.

Designed by Sir Herbert Austin himself, the Austin Seven was a true car in miniature and could seat two adults and two children. With the price reduced to £165 for 1923 – about forty-six weeks wages for the average worker in the motor industry – it was soon a success. This was only about £20 more than a cyclecar or large motor cycle and sidecar yet the Austin Seven offered more room and greater comfort in the same road space.

The Austin Seven did not just divert motor cyclists to cars but also created a new market for small cars. As the pioneering motoring journalist S F Edge said in 1925, 'Austin's case was an instance of that very uncommon phenomenon, a supply creating a demand, and filling it to the last ounce and every penny piece.'

Until 1924 there had been as many motor cycles in use as cars – about 400,000 of each. But the 1920s saw vehicle prices fall in real terms by more than fifty per cent as makers tried to increase demand and exploit the cost savings of mass production. By 1934 there were about 1,300,000 cars in use. Cars were now twice as common as motor cycles, and the number of motor cycles in use was going down.

The car in the Science Museum was presented by the Austin Motor Company Ltd in 1952. It is the second prototype and differs in many small ways from production cars. It was used for road testing by journalists and to illustrate advertisements and catalogues under the slogan 'The Motor for the Millions'.

The simple steel chassis was robust but not very

LIFE WORTH LIVING.
With an "Austin Seven" you can spend it where you will, for travel costs you about 1d. per mile with your jolly young family. The "Austin Seven" makes happiness inevitable, and it is a smart little turnout to be proud of.

A page from the Austin Seven sales brochure c 1922, featuring the car shown opposite

rigid. The arrangement of front and rear axles was a cheap and simple system but in practice gave somewhat uncertain handling. Cable-operated brakes were fitted on all four wheels but they were not very effective. The front brakes were controlled by a hand lever and the rear by a pedal. *The Light Car and Cyclecar* reported favourably, 'One is impressed by the silence of the axle, transmission and engine in top gear, whilst the suppleness of the suspension and the manner in which the car holds the road on corners are equally remarkable … The steering, too, is finger-light …', but they did concede that the brakes were 'not startlingly sudden and tremendously powerful in action, but rather retard gradually and progressively.' The 696 cc engine (increased to 747 cc for production cars) was similar in size to those in large motor cycles but built like much larger car engines. Its ten brake horsepower gave a top speed of about 50 mph in a car weighing only six and a half hundredweight. Fuel consumption was around 45 mpg.

Although electric lights are fitted, this prototype has no dynamo to power them. Electric starters became available on production cars in December 1923, speedometers in February 1924, and shock absorbers in March 1924. The basic design remained in production until 1938 by which time some 291,000 had been produced.

An affectionate, if rather robust, retrospective assessment in *Autocar* in 1967 said, 'In view of its haphazard steering and hopeless brakes I believe that a special providence watched over them, as they bumbled and rumbled and bounced and hopped over the highways of the nineteen-twenties.' But they offered a unique balance of comfort, performance and price which created and defined a new small car market.   PRM

LOGIE BAIRD'S TELEVISION APPARATUS

In a darkened room in January 1926 a group of men are peering intently at a dimly flickering pinky-orange oblong of light, about the size of a narrow credit card. Is there anything there? Can they truly see a barely recognizable human head and shoulders, features just discernible in the gloomy changing light? Is it real or is it imaginary?

The eccentric thirty-eight-year-old Scots inventor, John Logie Baird (1888–1946), presented early observers of television with a very crude picture. Yet his cumbersome mechanical system, skilfully marketed and publicized, generated a demand for television at a time when many thought it had little to offer.

Like all cinema film and television Baird's system relied on 'persistence of vision': so long as the small parts of a large picture are presented, in turn, quickly enough, the human eye can be fooled into seeing the whole without registering either individual elements or flicker. Thus a picture, scanned and broken into parts in one place, can be transmitted as an electrical signal to another location where, received and reassembled, it can be viewed as a recreated whole.

In his first demonstrations, Baird used a mechanical device to scan his subject, a puppet called Stooky Bill that can be seen in the main picture. The heart of the device was a large spinning cardboard disc, containing lenses arranged in a spiral. Baird's version of this idea, first suggested by Paul Nipkow (1860–1940) in Berlin in 1884, had thirty lenses. As the disc rotated, each lens scanned a different part of the subject. Light from each part of the subject fell, in turn, on to a light-sensitive electrical cell and was converted into an electrical signal. By the scanning process the pattern of light and dark in the original image was converted into a stream of electrical signals which could be sent along a wire or transmitted. Baird's system used the then common medium-wave radio band.

In the receiver the incoming signal caused a neon light bulb to flicker. A second Nipkow disc, spinning in time with the first, converted the flickers back into

lines. The watchers in 1926 saw these lines as a pinkish, glowing picture.

Although Baird's company claimed that the transmitter now in the Science Museum was his original apparatus, we now believe it to be a composite, containing some original parts, presented to the Museum in 1926 as part of a publicity drive. In collecting television-related objects, the Museum was responding to growing public interest. News of early experiments in the new medium had been published in the popular wireless enthusiast press of the time. Publicity about Baird's demonstrations, first in local, then national newspapers, had alerted the general public. Baird's company hoped the campaign would generate support for their attempt to persuade the Postmaster General to issue a licence for experimental public trials and to force the BBC to carry such broadcasts.

The campaign succeeded in raising public awareness, even though the generally acknowledged poor quality of the images discouraged the BBC and enraged other scientists in the television field, in particular Alan Archibald Campbell Swinton (1863–1930). He had suggested an electronic television

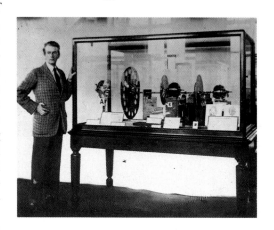

John Logie Baird presenting the 'transmitting portion of the original experimental Baird television apparatus' to the Science Museum, September 1926

system back in 1911 and objected strongly to what he saw as attempts to defraud the public, both through the claims that Baird had invented television and through the suggestion that the mechanical system was the only, or best, solution. Nevertheless, the company's persistence paid off. The BBC transmitted experimental television, using the Baird system, during the Depression years of 1929–35. The initial audience was small. Baird himself estimated that no more than thirty sets existed, twenty of which he believed had been home-built by enthusiasts.

Amazingly, despite the poor images and the constraints imposed on programme makers (head and shoulder shots only) public demand was generated. To satisfy it other workers, based in the mainstream electronics industry and dissatisfied with the limitations of mechanical scanners, turned their attention to electronic means of producing pictures. Baird, however, stuck stubbornly to his mechanical system, struggling to improve the quality.

In 1936 the world's first regular high-definition television service started. From the BBC's Alexandra Palace studio both Baird's mechanical system and the rival electronic system developed by EMI (See EMITRON CAMERA TUBE) were broadcast to the waiting public. Although much improved on the early apparatus the still unsatisfactory mechanical system was dropped after three months and the electronic EMI version adopted. Baird had not succeeded.

So why is Baird seen as the 'father' of television? It is probably because, rather than making any lasting technical breakthrough, he did more than any of his contemporaries to publicize television. He struggled gamely with an ultimately inadequate system whilst others in Britain, the United States, the Soviet Union and Europe contributed to the development of electronic television. Perhaps his fame owes something to the admiration the British traditionally lavish upon eccentrics. He is an example of the appealing underdog who holds that place in public affection traditionally given to the gallant loser.   JB

SUPERMARINE S.6B FLOATPLANE

On 13 September 1931 this Supermarine s.6B float-plane won the twelfth Schneider Trophy contest, capturing the Trophy for all time for Great Britain. Sixteen days later the same aircraft took the world air speed record to over 400 mph.

The Schneider Trophy contest was established in 1912 by Jacques Schneider with the aim of encouraging seaplane development. It was an international competition, with speed trials and tests of seaworthiness. The contest soon became one of the most prestigious aviation events, and with keen competition the Trophy passed between teams from France, Britain, the United States and Italy. By 1925 the British government were sufficiently convinced of the value of the contest as a stimulus to development that they undertook to support manufacturers in the design and construction of aircraft and engines for the competition. In subsequent contests pilots were also provided, drawn from the High Speed Flight of the Royal Air Force.

The s.6B had its origin in the s.4 streamlined twin-float monoplane designed by R J Mitchell (1895–1937) of the Supermarine Aviation Works in Southampton for the 1925 Schneider contest. Unfortunately this crashed during trials, but not before revealing the potential of the basic design. This was borne out in the 1927 contest, in which two s.5 floatplanes took first and second place.

By 1929 the Air Ministry considered that the Napier Lion engine, which had been fitted to all the previous Supermarine aircraft entered in the contest, had reached the limit of its development. In the search for another engine the Ministry chose a twelve-cylinder design from Rolls-Royce, which Henry (later Sir Henry) Royce undertook to develop for competition use. Mitchell designed the s.6 floatplane to take this 1,900 horsepower engine, which went on to win the 1929 contest. Under the rules, a third win would enable Britain to take permanent possession of the Schneider Trophy.

British participation in the 1931 contest remained

Refuelling at Calshot, 1931

in doubt until early that year. The government was unwilling, in the light of the prevailing economic depression, to provide support. In the event Lady Houston, a wealthy and fervently patriotic widow, gave the £100,000 necessary for a British team to defend the Trophy. Too little time was by then left for Supermarine and Rolls-Royce to produce new designs, so work started on developing the existing airframe and engine as much as possible.

The resulting aircraft was designated s.6B. Like its predecessor it was an all-metal monoplane mounted on twin floats. Much of the aluminium surface of the aircraft was double-skinned to provide drag-free cooling radiators for water and oil. Two s.6B aircraft were built to this design, and were given numbers s1595 and s1596.

In a gruelling development programme the power output of the Rolls-Royce R engine was raised to 2,330 horsepower, both by increasing engine speed and improving the supercharging. The connecting rods, crankcase and crankshaft were all re-designed to withstand greater stress, and a special fuel devised for race conditions.

By 12 September 1931, the date set for the contest, both the French and Italians had withdrawn because of developmental difficulties with their aircraft, and

so, after a one day postponement owing to bad weather, s1595 piloted by Flight Lieutenant John Boothman flew round the course at Calshot at an average speed of 340.08 mph to win the Trophy uncontested. Later that day the world air speed record was raised to 379.05 mph by s1596, flown by Flight Lieutenant George Stainforth. Rolls-Royce were keen to better this, and so on 29 September 1931 s1595, powered by a 'sprint' version of the R engine using another special fuel and producing 2,530 horsepower, took the world air speed record to 407.5 mph with Stainforth at the controls.

The experience gained by both Supermarine and Rolls-Royce in achieving these results provided an important basis for their subsequent development of high speed aircraft and engines. Mitchell's design of the Spitfire fighter aircraft grew directly from this work for the Schneider Trophy aircraft. Similarly, the Rolls-Royce Merlin engine, the most widely used British aero engine in the Second World War, benefited greatly from the development of the R engine. Experience with the R engine supercharger also provided an important basis for the design of centrifugal compressors, vital both to Merlin engine development (See THE ROLLS-ROYCE MERLIN) and to the jet engine (See GLOSTER-WHITTLE E.28/39 JET AIRCRAFT). Information was also gained on the importance of fuel quality to improved engine performance; and the use of tetra-ethyl lead to suppress detonation or 'knock' anticipated the provision of 100-octane petrol during the Battle of Britain – an important element in the performance of the Spitfire and Hurricane. Thus the contribution of the Schneider Trophy contest to the Allied war effort proved to be of great value.

Late in 1931, s.6B s1595 was placed on display in the Science Museum, followed in 1948 by the Rolls-Royce R 'sprint' engine used in the second air speed record attempt. Finally the Schneider Trophy itself arrived in the Museum in 1977, joining the aircraft whose creation it had so notably inspired. AJH

LAWRENCE'S ELEVEN-INCH CYCLOTRON

'I'm going to be famous!', the young experimenter yelled as he ran exuberantly across the university car park. Ernest Lawrence had not observed something new; he had conceived a brilliant idea that, he knew, would make it possible to split the atom much more easily than other scientists had dared to believe. Lawrence went on to build his dream – the cyclotron – enabling atoms to be split in apparatus that could be mounted on a laboratory bench.

Lawrence (1901–58) was twenty-seven when he startled his colleagues by moving from Yale to the University of California at Berkeley, then a little-known state university. He soon took up the challenge of probing atomic nuclei, about which little was known except that they are positively charged and that they measure only about a billionth of a centimetre across. The best hope of investigating nuclei more closely was simply to see what happens when they are bombarded with *other* nuclei, such as protons (hydrogen nuclei). But, in order to arrange such collisions, these particles had to be accelerated to high energies so that they could actually penetrate the target nuclei. This acceleration can, in principle, be achieved by using a very high electrical voltage to increase the energy of the particles as they move in a straight line. In practice, several hundred thousand volts are necessary, and this can result in severe problems associated with electrical breakdown caused by sparking.

One evening in the spring of 1929, Lawrence came up with an ingenious solution to this problem. He was in the library at Berkeley trying, with his rudimentary German, to read a paper on particle accelerators, *Über ein neues Prinzip zur Herstellung hoher Spannungen*, by Rolf Wideröe. Inspired by what he read, he suggested that instead of accelerating particles in a straight line, it may be that they could be whirled round in a spiral motion, gradually gaining energy.

Lawrence soon developed his first notions into a new type of accelerator design. The paths of the particles emerging from the centre of the apparatus would be curved by the magnets located on both sides of the two hollow cavities (later called dees, after their shape). By applying a voltage across the gap separating the dees, the particles could be accelerated as they cross it for the first time. When the particles next pass the gap, they would be travelling in the opposite direction, but again they could be accelerated by judiciously reversing the direction of the voltage. By continually repeating this process, the particles could be accelerated to high energies, ready to be projected towards target nuclei. As Lawrence showed mathematically, the time taken for each half-revolution is the same regardless of the diameter of a particle's orbit so, remarkably, the reversal of voltage works equally well for all the particles in the apparatus.

Lawrence believed he had struck gold but some of his colleagues were sceptical that the idea could be

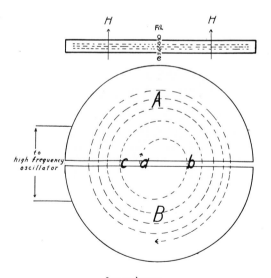

In a cyclotron,
charged particles are accelerated in a spiral path
between two large magnets

made to work, and this cool reception probably took the wind out of his sails. For someone of Lawrence's drive and ambition, it is perhaps surprising that he took nearly two years to bring his initial plans to fruition. In April 1931, he and his student M Stanley Livingston reported the first working cyclotron which was made by using a magnet of four and a half inches in diameter and which produced protons that had, in effect, been accelerated by 80,000 volts. In his laboratory ten months later, Lawrence danced with glee when he saw that his new eleven-inch cyclotron could accelerate protons through more than a million volts.

In his zeal to improve the design, he had fallen behind his competitors and lost the race to split the atom, which was won by Cockcroft and Walton at the Cavendish Laboratory in Cambridge (See COCK-CROFT AND WALTON'S ACCELERATOR). In June 1932, they observed an astonishing reaction in which a nucleus was split by an incoming proton into two helium nuclei, a result that Lawrence and his colleagues reproduced three months later. In this case, it was the runner-up who won first prize: Lawrence was awarded the Nobel Prize for Physics in 1939, twelve years before Cockcroft and Walton were awarded theirs.

Cyclotrons were later used to do many experiments to probe the structure of the atomic nucleus and to discover new chemical elements, several of which were discovered at Berkeley. They are now used to produce radioactive materials for use in medical diagnostics and for treating diseases such as cancer. However, it is in the field of pure science that Lawrence's original idea has had most applications. It is the basis of the design of all of today's particle accelerators which allow scientists to probe the very structure of matter, into regions even smaller than the nucleus. Lawrence's first reaction to his brainwave was quite right: he had indeed conceived an idea great enough to make him famous.   GPF

COCKCROFT AND WALTON'S ACCELERATOR

In the 1930s nuclear physics became a distinct speciality within physics as rapid progress was made in understanding the atomic nucleus. While major discoveries such as the neutron, the positive electron, and artificial radioactivity (*See* CHADWICK'S PARAFFIN WAX and DISCOVERY OF ARTIFICIAL RADIOACTIVITY) helped to establish the new field, one of the most important factors was the construction and use of new machines for experimenting on nuclei. These were known as particle accelerators or atom-smashers. Two became operational in 1932 and the Science Museum has parts of them; one built by John (later Sir John) Cockcroft (1897–1967) and E T S Walton (b. 1903) at the Cavendish Laboratory in Cambridge and the other the eleven-inch cyclotron built by Lawrence and Livingston in Berkeley, California (*See* LAWRENCE'S ELEVEN-INCH CYCLOTRON).

Though these machines had important differences, they both needed space and teams of people to operate them, and so helped usher in the 'big science' that was such a distinctive feature of research after the Second World War. This trend has continued, and a modern accelerator, such as LEP at CERN (European Organization for Nuclear Research) near Geneva, while still incorporating features of these original machines, is physically more than 10,000 times larger and employs hundreds of people to operate and maintain it.

The accelerator built by Cockcroft and Walton was the first to be used for experiments on nuclear physics and helped to keep the Cavendish Laboratory in the forefront of research in nuclear physics in the early 1930s. In this machine, protons, the nuclei of hydrogen atoms, are released at the top of a glass column which has been emptied of air. The protons have a positive electrical charge and as they travel down the glass column they pass through a series of metal cylinders that are electrically charged. The effect of this arrangement is to accelerate the protons. At the far end of the glass tube, placed in the way of the protons, is the target: a piece of metal or

Cockcroft and Walton's accelerator (The Cavendish Laboratory, University of Cambridge)

other material. The protons collide with the nuclei of atoms in the target and break them into fragments. By examining the fragments, physicists were able to find out more about the detailed structure of these nuclei.

By 1930 physicists were keen to develop accelerators for several reasons and by then they had the advantage of high-voltage equipment developed by the electrical industry. From before the First World War, when Ernest Rutherford (1871–1937) had first suggested the atom had a central nucleus, most experiments on the nucleus involved radioactive materials. Rutherford, for example, often used polonium as a source of alpha particles (helium nuclei) for bombarding the nuclei in a target. However, these radioactive materials were both expensive and awkward to use. In addition, the particles they emitted had a range of energies so it was difficult to

interpret the results. The new accelerators were designed to get round these difficulties by providing beams of particles with one energy.

With Rutherford's encouragement there were various attempts to build accelerators at the Cavendish in the 1920s. But these were not successful until Cockcroft and Walton pooled their resources. To build their machine, they took advantage of some new technology. Before going to Cambridge, Cockcroft had been an apprentice with Metropolitan-Vickers in Manchester. They made transformers and other items used in the transmission of electricity. That experience and his continuing contact with Metropolitan-Vickers were very important for Cockcroft when he came to build a machine to accelerate particles using high voltages. One particular example of the value of the Metropolitan-Vickers connection was that the glass tubes (originally designed for petrol pumps) and other components of the accelerator were sealed using special compounds which had only recently been developed by C R Burch who worked there.

The first experiments by Cockcroft and Walton involved bombarding lithium atoms with protons, splitting them to make helium nuclei. For this work they were awarded the Nobel Prize for Physics in 1951. Very quickly their original machine was superseded by a larger version built by the Dutch electrical firm Philips. Even today the principle of the Cockcroft-Walton machine is often used in the first stage of the largest accelerators, to accelerate the particles so that they can be injected into a circular cavity where they can be given even higher energies. These latest machines are used to search for evidence for quarks and other exotic particles that, according to current theories, are thought to make up the more familiar protons and neutrons.

The Science Museum was presented with a portion of Cockcroft and Walton's original apparatus for the artificial disintegration of the elements by swift protons by the Cavendish Laboratory in 1933.   AQM

CHADWICK'S PARAFFIN WAX

These unprepossessing pieces of paraffin wax were used by James Chadwick (1891–1974) early in 1932 in experiments which established the existence of neutrons, particles found alongside protons in the nuclei of atoms.

Following the announcement of this discovery in the journal *Nature*, a journalist from *The Times* reported Chadwick's own assessment of the significance of his discovery as follows: 'Positive results in the search for "neutrons" would add considerably to the existing knowledge on the subject of the construction of matter, and as such would be of the greatest interest to science, but, to humanity in general, the ultimate success or otherwise of the experiments that were being carried out in this direction would make no difference.'

However, Chadwick's off-the-cuff remarks about the significance of his discovery turned out to be remarkably wide of the mark since it led to the development of atomic weapons within thirteen years. In the short term, however, his assessment was correct; in 1932 the neutron was important only for the small number of scientists interested in the structure of the nucleus.

Chadwick's own views about nuclear structure were strongly influenced by ideas put forward by Rutherford a decade or so earlier, shortly after the latter had left Manchester to become Cavendish Professor at Cambridge. At that time it was thought that the atom was made up of a nucleus surrounded by negatively charged electrons. Rutherford specu-lated that if the nucleus were made up of positively charged particles, for which he suggested the name protons, and extra electrons in addition to those found outside the nucleus, then it might be possible for a proton to become closely bound to a nuclear electron in a unit he termed the 'neutron'. Over the next few years experiments were done at the Cavendish to detect these neutrons without any success.

During this time substantial progress was made in solving problems about the behaviour of the electrons orbiting outside the nucleus. But as quickly as problems were solved for the electrons outside the nucleus, the new theories created problems for the electrons inside it. At the end of 1930, in a desperate attempt to solve some of these mysteries concerning nuclear electrons, Wolfgang Pauli suggested another kind of neutron for the nucleus, a particle much lighter than the version proposed by Rutherford.

This theoretical impasse was broken by new experimental evidence. In Germany, France and Britain physicists had used alpha particles (helium nuclei) to bombard samples such as beryllium. In these experiments a nucleus weighing four units was made to collide with one weighing nine units, and the effects were quite dramatic. The results were puzzling but suggested a new effect of radiation. However, Chadwick did not agree with this conclusion and carried out his own experiments. He put these pieces of paraffin wax in the way of the mysterious radiation from beryllium. As other experimenters had found, the radiation knocked hydrogen nuclei (ie protons) from the wax, which could easily be detected because they have an electric charge. Chadwick then repeated the experiments using nitrogen and measured the recoil of the nitrogen nuclei. From his measurements of the energy of the protons and the nitrogen nuclei Chadwick concluded that they had been hit by particles with the same mass as the proton but which were electrically neutral – in other words Rutherford's neutron. This was the view of the neutron he put forward in the papers announcing his discovery.

Within a few months it was generally accepted that nuclei were now made up of protons and neutrons while nuclear electrons went out of favour. Pauli's particle also lost its place as a constituent of the nucleus but it too was transformed and became a particle created in various nuclear processes. Fermi suggested the name 'neutrino', to distinguish this particle from its heavier namesake.

Chadwick's initial estimate of the value of his discovery was accurate at the time. However, the property of neutrons to split uranium nuclei releasing large amounts of energy was discovered in the winter of 1938–9 a few months before the outbreak of the Second World War. By the end of the war, less than six years later, and only thirteen years after Chadwick's discovery, neutrons had been used for chain reactions in nuclear reactors and nuclear weapons. Contrary to Chadwick's view, the neutron was a discovery that had a profound effect on all humanity in a remarkably short time.   AQM

Opposite: The circular paraffin wax
samples with pieces of metal foil also used
in Chadwick's experiments

THE DISCOVERY OF POLYETHYLENE

In 1932 the unambiguously titled book *Chemistry Triumphant* announced that we were destined to live in 'The Silicon-Plastic Age'. While silicon was temporarily forgotten, the term 'Plastic Age' has proved a suitable title for the mid-twentieth century. Of course at that time the new materials were in their infancy. Although the flammable celluloid and a selection of rather brittle plastics were available, Bakelite being perhaps the best remembered (*See* THE FIRST PLASTIC), the generation of new materials with which we are familiar was only just being developed. Commercially produced nylon, polystyrene, perspex and polyvinyl chloride (PVC) were all children of the 1930s. Indeed the plastic destined to be made in the largest quantity of all, polyethylene (better known as Polythene, ICI's trade name for the new product), was only developed in 1933, the year after the plastics age had been baptized.

The Brunner Mond Company, which with Nobel Industries had taken the lead in forming ICI in 1926, had become ICI's Alkali Division. Infused with the tradition of Ludwig Mond, the Alkali Division opened the new Winnington Research Laboratory in 1928, with many newly recruited scientists, and embarked on an ambitious programme of fundamental research – and all this during a world-wide recession. ICI acquired apparatus from Anton Michels, a Dutch physicist who had devised apparatus for submitting materials to pressures up to 2,000 atmospheres (atm), and set up a research programme hoping to find that high pressures would produce new and potentially useful effects.

The scientist first assigned to this work in 1931 was Reginald Gibson, who had worked with Michels in Amsterdam; a year later he was joined by Eric Fawcett, an organic chemist recruited to work on ICI's ambitious plan to process oil from coal (a project that had become a casualty of the recession).

After much fruitless work, they set up an experiment on Friday 24 March 1933 in which a mixture of ethylene and benzaldehyde was heated to 170 degrees centigrade, the pressure was raised to 1,700 atm and the apparatus was left in the hope that the two substances would combine together. The experiment had been suggested by Professor (later Sir) Robert Robinson to test the idea that the normally slow 'Diels Alder' reaction might go rapidly, and without a catalyst, under pressure.

On Monday no evidence of the desired reaction was found, but Gibson wrote in his notebook, 'waxy solid found in the reaction tube'. Fawcett demonstrated that this was a polymer of ethylene, which could be melted and drawn into threads.

Part of the laboratory where polyethylene was discovered (ICI)

Attempts to repeat the experiment were usually unsuccessful: either nothing happened, or the reactants decomposed explosively. It should be emphasized that this early apparatus was amazingly crude, a manually operated pressure pump, with a reaction vessel in a bath of hot oil. At the time it was generally thought that it was impossible to make solid polymers of ethylene, so when in September 1935 Fawcett commented at a conference in Cambridge that he had done so he simply wasn't believed. This was fortunate for ICI because there was no patent application on file. Only in December of that year, using greatly improved equipment, was Michael Perrin able to carry out the polymerization reproducibly and establish the basis of a production process. Following this work progress was rapid; in 1937 a ten-ton per year pilot plant opened, to be followed by a 100-ton per year plant which started up on the first day of the Second World War.

Although we perhaps associate polyethylene with disposable packaging, its first uses were seen to be for insulation, and almost immediately this became of crucial importance. The 1,000-ton plant approved in 1940 would provide the lightweight compact insulators that made possible the air- and ship-borne radar that enabled Britain to win the Battle of the Atlantic (*See* RANDALL AND BOOT'S CAVITY MAGNETRON). No longer was polyethylene an interesting technological oddity, but a war-winning weapon.

The apparatus used by Gibson and Fawcett was presented by ICI to the Science Museum on the 50th anniversary of the first experiment in the discovery of polyethylene.

RFB

Advice from Dr Anthony Willbourn

THE BOEING 247D

Occasionally, technological change occurs so rapidly that it becomes inextricably linked with a particular object, and over the years that object comes to stand for both the advance and the social changes associated with it. Just as the rise of railways and the growth in passenger traffic is associated in the minds of many with Stephenson's *Rocket*, the explosive growth in the number of air passengers and the establishment of regular air routes over much of the globe during the 1930s is automatically linked with the Douglas DC-3. But on reflection, perhaps we should award the title of 'the world's first modern airliner' not to the DC-3, but to the sturdy, purposeful Boeing 247D. An example is preserved by the Science Museum in the National Aeronautical Collection.

In the late 1920s and early 1930s the United States of America was entering a 'golden age' of aeronautical design; on the drawing boards of companies such as Boeing and Douglas were studies which would lead to advances such as metal construction, variable pitch propellers, the pressurized airliner cabin and the retractable undercarriage. The establishment of government sponsored air mail routes across the nation, initially serviced by ex-World War One biplanes, had led several manufacturers to produce fast, metal monoplanes and these aircraft were used by the emerging airlines to compete for the lucrative mail contracts. The Boeing 'Monomail' was the company's answer to the problem, developed to carry not just 750 pounds of mail but six passengers in a separate compartment. Eventually, these aerodynamic and structural advantages were brought together in the Boeing 247. This revolutionary aircraft had most of the attributes associated with the modern airliner and the prototype first took to the air in February 1933.

With the ability to climb safely on the power of only one of the two engines, the 247's advanced specification, including controllable pitch propellers, retractable undercarriage and wing and tail de-icing virtually rendered all other airliners obsolete over-

night. Boeing at this time was involved in both airline ownership and management, in that the United Air Lines system and its subsidiaries were part of the company. This confusion between producer and consumer (which had been experienced briefly by Handley Page in Great Britain) caused the developed version of the airliner, the Boeing 247D, to be designed for the United Air Lines route structure. This meant that although over seventy aircraft were eventually built, only UAL used them in quantity. Boeing's withdrawal from the airline ownership business in 1934 allowed UAL to buy the newer Douglas airliners, so that by 1938 only twenty-five 247D were left in their fleet, whilst the DC-3 'Mainliners' were operating the prestigious fifteen-hour coast-to-coast service.

The same qualities of reliability and a respectable range, caused one of the Boeing airliners to lead a very exciting life: NR-257Y was selected by Colonel Roscoe Turner, the flamboyant American air racer, and Clyde Pangbourne for the 1934 MacRobertson air race from Mildenhall in Suffolk to Melbourne, Australia. Nursing overheated engines, Turner and Pangbourne came in a creditable third behind the eventual winners, Scott and Campbell Black in a de Havilland 88 Comet. It was highly significant that a KLM DC-2 carrying paying passengers, actually came

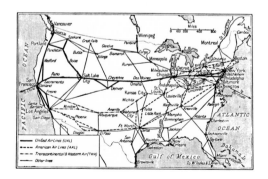

The 'Chief Air Lines of the United States' from Robert Finch, *The World's Airways*, 1938

in ahead of the 247D, just twenty hours behind the winning Comet racer; this performance was so impressive that the decline of the Boeing 247, and the rise of the DC-2/DC-3 line can be traced to this event.

It might be thought that the outbreak of World War Two would see a new lease of life for the Boeing design, as it did with many other transport aircraft. However, the limited number of 247s that were requisitioned by the US Army Air Corps suffered from the disadvantage of having main wing spars that penetrated the cabin, and this restricted their usefulness in carrying cargo. This may account for the fact that the aircraft were declared surplus to requirements in 1944, and sold off, mainly to private owners.

Although coming to the end of their service lives, one more aircraft of this type was to play a powerful role in the history of aviation. The Royal Air Force had obtained a 247D, from the Royal Canadian Air Force, and this aircraft, serialled DZ 203, was used in the development of centimetric radar (*See* RANDALL AND BOOT'S CAVITY MAGNETRON). This 247's main claim to fame, however, was that it performed one of the world's first fully automatic approach and landings, on 10 February 1945 at RAF Defford in Worcestershire. The aircraft was flown by Group Captain Frank Griffiths and was equipped with a modified Minneapolis Honeywell Autopilot. Regrettably, the aircraft was destroyed when a large oak tree was blown down on to the hangar it was occupying at Defford.

The world population of Boeing 247s is now very small; although the Science Museum's example had a fairly unremarkable career with various private owners after World War Two, it became famous on its very last flight. Purchased after the collapse of the 'Wings & Wheels Museum' in Florida and flown across the old World War Two 'Northern Ferry Route', when N18E touched down at Wroughton in Wiltshire on 3 August 1983, it became the oldest aircraft to have flown the Atlantic.   GRS

DISCOVERY OF ARTIFICIAL RADIOACTIVITY

Among the remarkable developments in nuclear physics during the early 1930s was the discovery of artificial radioactivity. Appropriately, it was made by Irène Curie (1897–1956) and Frédéric Joliot (1900–58), the daughter and son-in-law of Marie Curie who had made important contributions to the study of radioactivity at the turn of the century. Marie Curie had studied naturally occurring radioactive materials: it fell to the next generation to investigate processes by which materials could become radioactive.

In 1932 the Joliot-Curies were involved in experiments leading to the discovery of the neutron by James Chadwick, (See CHADWICK'S PARAFFIN WAX) and were unlucky not to make the discovery themselves. The pace quickened later that year with Anderson in California finding a new positive particle. The evidence for this particle came from photographs of cosmic rays (the debris produced as particles from space travel through the atmosphere) passing through a lead plate. Very soon Anderson's discovery was linked to a prediction made by Paul Dirac in Cambridge of a positively charged version of the electron and Anderson's particle became known as the positron. These discoveries, together with Joliot and Curie's work on artificial radioactivity, laid the foundations for nuclear physics, which by the mid-1930s had become a distinct speciality within physics.

The significance of these separate discoveries became clear in 1933 when Curie and Joliot continued the experiments that had provided Chadwick with an important clue in his discovery of the neutron. Samples of elements like boron were bombarded with alpha particles (helium nuclei) emitted by polonium (an element discovered by Marie Curie). But now Curie and Joliot noticed that positrons were also emitted in some cases. More remarkable was that when the source of alpha particles was removed so that no more nuclear collisions took place, these positrons were still detected. They had discovered a new nuclear process, analogous to the emission of an electron by the nucleus, the process of beta-decay discovered at the turn of the century. In beta-decay, nuclei do not disintegrate immediately but only after a time; this varies from nucleus to nucleus because the decay is subject to chance.

To confirm that they had discovered a new kind of

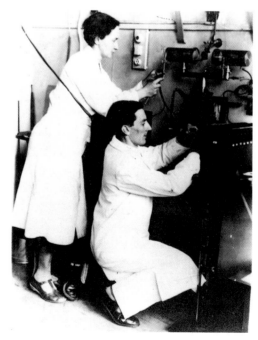

Top: Sketches by Frédéric Joliot describing the apparatus used for producing artificial radioactivity. Above: Irène Curie and Frédéric Joliot at work in their laboratory, c 1934 (Institut Curie)

radioactivity, Curie and Joliot hit on an ingenious experiment to prove that it was the nucleus that was involved. They took a small U-shaped tube and filled one arm with a boron compound and sealed the tube. The boron sample was bombarded with alpha particles (helium nuclei) to convert some boron nuclei into nuclei of nitrogen. According to Joliot and Curie, these nitrogen nuclei would now emit positrons. The clever twist to their experiment was to separate this nitrogen from boron or anything else that might be present and to show that it was only nitrogen that emitted positrons. What they did was to collect the nitrogen in the form of ammonia in the other limb of the tube. By the end, all the radioactivity had left the boron compound and was carried with the nitrogen. This showed that the boron had been converted to nitrogen. In 1935 they were awarded the Nobel Prize for their work on this new kind of radioactivity.

Another interesting aspect of the experiment was that Curie and Joliot used electronic methods of counting the positrons they detected. These new methods were developed in the early 1930s and often used components designed for radio receivers.

The discovery of the neutron and of new types of radioactivity in the early 1930s paved the way for many new experiments on the properties of different nuclei. Towards the end of the 1930s, Enrico Fermi and others in Rome, and Otto Hahn and Lise Meitner in Berlin carried out experiments bombarding heavy nuclei like uranium with neutrons. This led to the discovery of nuclear fission a few months before the outbreak of the Second World War and so to the race to build the first atomic weapons. What had been an esoteric branch of science involving a few people such as Marie Curie and Ernest Rutherford, in one generation had become important enough to change the course of history.   AQM

REYNOLDS' CINERADIOGRAPHY APPARATUS

The public first marvelled at the tentative beginnings of cinema in Paris in 1895. It was also the year in which a then unknown German professor, Wilhelm Conrad Röntgen (1845–1923), first described X-rays, a different but equally startling means of photography. By 1897, investigators were already seeking ways in which the two new technologies could be merged. What they sought was cineradiography – moving X-ray records of the body's innermost workings: the heart as it pumped, the stomach as food passed through, the joints as they moved.

The apparatus that Röntgen had used to produce the first X-ray photographs in late 1895 had been quickly copied by physicists, photographers and doctors in many countries and X-rays had found immediate practical application in medical matters. In 1896, for example, the physicist A A Campbell Swinton succeeded in localizing a bullet lodged in the head of an unfortunate patient. All that was required to take X-ray photographs was an electrical circuit incorporating a power source, such as a battery, a means of increasing the voltage and a special glass 'discharge' tube, containing a partial vacuum and with electrodes sealed into each end. When current passed, this tube emitted X-rays. Placing an object between the tube and an ordinary photographic plate produced the X-ray image. The rays easily penetrated flesh, showing bone and other denser structures as shadows.

One such amateur X-ray set was put together in 1896–7 by a London general practitioner, John Reynolds, with his fifteen year old son, Russell. Ten years later Russell Reynolds (1880–1964) qualified as a doctor and from 1922 devoted his professional research to the much more challenging technical problem of cineradiography. The cineradiographic apparatus in the Science Museum represents a high point in Reynolds' career. It was the first commer-cially produced outfit for cineradiography in Britain, designed to his specifications in 1935 by Watson & Sons, the foremost British manufacturer of radiological equipment. Reynolds himself owned this set and it seems that only one other example was ever produced.

The technical problems Reynolds faced were enormous. From about 1900, workers trying to make true cineradiographic films had used one of two methods. The 'direct' method recorded the X-ray images directly on to cinefilm sandwiched between two fluorescent screens. These reinforced with visible light the action of the X-rays on the cinefilm. However, X-rays could not be focused so film sizes were huge; to record adult chest movements, for example, a film strip 120 feet × 15 inches wide was required. A few research films of small organs such as the gall bladder were made, but for practical purposes the method was useless. Other workers, including Russell Reynolds, pursued the indirect method – that is, they attempted to film the X-ray images that could be produced on a fluorescent screen, using a cine-camera. The main problem here was to get a bright enough image. The high doses of radiation required were potentially harmful to the patient and often put an intolerable strain on the X-ray tube.

By 1925 Reynolds had partially solved the problem. He incorporated a switch which synchronized the production of X-rays with the opening of the cine-camera shutter, thus halving the risk to patient and tube. By 1934, he had obtained cinefilms of almost all the joints at speeds of up to twelve frames per second and was able to convince Watson & Sons that the method was both practical enough and potentially useful enough to warrant commercial production. Cautiously, and perhaps wisely, Watsons put together much of this first cineradiography out-fit by modifying existing equipment lines. Cineradiography was never to be a money spinner. Research interest in the technique remained high during the 1940s and '50s and teaching films were made showing, for example, the working of the iron lung in artificially respirated patients. Reynolds himself achieved cinefilm of moving joints at up to fifty frames per second and 'slow motion' film, essential to understanding the heartbeat, by projecting film he had made at twenty-five frames per second. Early barium meal examinations, which made the intestinal contents radio-opaque, were also recorded. Reynolds achieved these later successes not by any single breakthrough but by constant modifications in technique, exploiting to the full every new improvement in the technologies concerned; wider aperture lenses became available, as did more brilliant fluorescent screens and more sensitive film.

In 1951 Watsons brought out a second model of Russell Reynolds' outfit, incorporating these improvements. This was exhibited at the Festival of Britain in 1951 as an outstanding example of British scientific equipment. However, only four examples were ever produced, three going to hospitals and one to a research institution. Cineradiography had yet to be incorporated into the routine work of the X-ray department. Incorporated it was, however, a decade later, largely due to the invention of the electronic image amplifier, which solved at a stroke most of the problems with which Reynolds had struggled for years. Watching and recording moving X-ray images on television monitors became both simple and routine, and is now an integral part of complex procedures such as cardiac catheterization. In Britain, it was the work of Russell Reynolds, under great technical difficulties, that had kept the prospect of successful cineradiography to the fore.

GML

ORIGINAL RADAR RECEIVER

On 26 February 1935, Robert Watson-Watt (1892–1973, knighted in 1942) carried out an experiment which showed that a distant aircraft could reflect radio waves transmitted towards it, and that sufficiently sensitive equipment with a cathode ray tube could detect the radio echo. The previous day, equipment from the Radio Research Station near Slough was loaded into an elderly Morris van and driven to a field at Weedon, Northamptonshire. The equipment was duly adjusted, and the next day Watson-Watt arrived with A P Rowe (Assistant to H E Wimperis, Director of Scientific Research to the Air Ministry) to join his own assistant, A F Wilkins. Secrecy surrounded the proceedings; Dyer, the van driver and labourer, and Patrick (Watson-Watt's nephew, who had gone along for a day out) had to wait outside the field. The pilot of a Heyford bomber had been ordered to fly to and fro at 5,000 feet in the powerful beam from the BBC's nearby Empire Short-Wave transmitter at Daventry. Convincing deflections of the spot on the screen were seen when the aircraft was some eight miles away, and Rowe, now satisfied that aircraft could be detected by radio, set off back to London in Watson-Watt's car to report back to the Ministry. They had reached Towcester when they realized that, in their excitement, they had forgotten Patrick, and had to turn back.

At this time, Watson-Watt was Superintendent of the Radio Division of the National Physical Laboratory. His work was concerned with understanding upper-atmosphere reflection – the basis of global radio communication in the pre-satellite era. Since Marconi's 1901 transatlantic demonstration, the power of radio for communication, entertainment and propaganda had set off several trains of investigation into the way radio waves followed the curvature of the Earth. In England, Sir Edward Appleton pioneered this work in 1924, pointing a radio beam skywards and analysing what came back; he was honoured by having one of the ionized layers of the upper atmosphere named after him.

One of the standard tools of the Radio Research

Station's investigations was a crude two-channel radio receiver, linked to a cathode ray tube which indicated the strength of reflections. It was one of these instruments which was used for the now-celebrated Daventry experiment. Watson-Watt's motive in mounting the experiment and showing it to A P Rowe dated back to a request he had received from the Air Ministry in 1934 to investigate the

Top: A 110-metre wartime radar transmitter tower;
Britain's 'Chain-Home' coastal-defence stations had
three or four, with aerial wires strung between them
(Imperial War Museum).
Above: Sir Robert Watson-Watt at the presentation
of the 'Daventry Experiment' receiver to the
Science Museum in 1957

feasibility of a 'death-ray' method of ensuring that bombers did not get through to their targets. His first calculations suggested that a lethal beam was not practicable, but late in January 1935 one of his memoranda concluded,

Meanwhile attention is being turned to the still difficult but less unpromising problem of radio-detection as opposed to radio-destruction...

Only weeks later (and before the important experiment of 26 February) Watson-Watt proposed the linking of radio-detection stations to a fighter-control network. On the basis of the Daventry demonstration, and the Air Ministry's reading of the worsening European situation, the development of coastal defence radar in England linked to a system of fighter control went ahead remarkably rapidly and paved the way for the British (and later, Allied) superiority in radar which was to be a decisive factor in victory during the Second World War. The 'who invented radar?' controversy is still running today, but no one denies Watson-Watt's central role in turning British radar from an experiment in an old van into a vital instrument of defence.

In 1951, Watson-Watt and some of his collaborators (who had formed a syndicate) became embroiled in an extraordinary tussle with the Royal Commission on Awards to Inventors. During the proceedings, the syndicate suggested that 'threepence [old pence] per head in respect of himself and of each of those dependants who are not able to speak for themselves' might not be an unreasonable sum to seek – yielding about £675,000. The actual award was £87,950. Watson-Watt's powers of persuasion failed in more light-hearted circumstances in 1954 in Canada. *En route* to a learned gathering to speak about his radar work, he was stopped for speeding by a highway patrol who were using an early radar speed trap. The officer remained unmoved by Watson-Watt's claims and demanded payment of a fine on the spot.   ED

MANCHESTER DIFFERENTIAL ANALYSER

Many problems in science and technology can be described by differential equations; in the 1920s and 1930s early analogue mechanical computers, 'differential analysers', were built to solve them. Modern digital computers are essentially counting devices which can handle numbers, but these differential analysers solved equations by modelling mathematical relationships with shafts and gears.

This reconstructed section of the Manchester University differential analyser of 1935 is one of the surviving relics of this era in computing. It is about half the size of the complete machine, built by the local engineering firm of Metropolitan-Vickers under the supervision of Professor D R Hartree (1897–1958). He first used it to solve problems in atomic physics, and during the Second World War it was used for military research, including work on the cavity magnetron (*See* RANDALL AND BOOT'S CAVITY MAGNETRON).

Its design was based on the first differential analyser, completed in 1930 at the Massachusetts Institute of Technology by a team led by Vannevar Bush (1890–1974), an electrical engineer who later marshalled America's scientific effort in World War Two, and also played a key role in the decision to drop the first atomic bomb.

The drive to develop the differential analyser came from the growing telecommunications and electricity supply networks of the 1920s. To design power transmission networks which would not be liable to blackouts, engineers had to solve complex differential equations which could take months of painstaking work. Bush realized that 'better ways of analysing were certainly needed'. His solution, like that of Charles Babbage a hundred years before, was to mechanize the process of calculation (*See* BABBAGE'S CALCULATING ENGINES). In the 1920s the only feasible way of making such a computer was to use mechanical rather than electrical technology. Relatively complex analogue mechanical computing devices had been developed to aim the guns aboard battleships, but Bush's machine was the first civilian analogue mechanical computer. The idea of the differential analyser had been put forward by the British physicist and entrepreneur Lord Kelvin in 1876 (he did not attempt to build one), but Bush's team did not know of Kelvin's ideas until after their differential analyser had been completed.

The Massachusetts Institute of Technology differential analyser was widely copied in Europe as well as America. It was the best means of solving differential equations until the development of the electronic computer in the 1940s. During the Second World War, differential analysers were chiefly used to compute firing tables for guns, as well as to design radar control systems and electronic circuits. However, setting up a differential analyser could take several days. Although hybrid, part-mechanical part-electrical differential analysers were built during the war to speed up the process, in the end, they were a dead-end technology, made obsolete by electronic digital computers.

Hartree wrote that 'my first impression on seeing the photographs of Dr Bush's machine was that they looked as if someone had been enjoying himself with a super-Meccano set'. Hartree began trying to build a Meccano model 'more for amusement than with any serious purpose', which was so successful that with the help of a student, Arthur Porter, he built a small differential analyser using many standard Meccano parts. It was capable of useful work, and gave good practice in 'programming' whilst the full-size analyser was under construction.

The differential analyser was a purely mechanical device – an 'analogue computer', which worked by representing each element of the equation by a rotating shaft. Gearing allowed the shafts to be set to rotate relative to each other, reproducing the relationship of elements of the equation. Experienced operators could follow the mathematics of solving an equation by watching the turning shafts.

The first step in using a differential analyser was to set up the shafts to reproduce the equation to be solved, a time-consuming operation. Graphs were prepared showing how changing variables in the equation altered with respect to one another, for example in the case of a train's speed changing with time. These graphs were placed on input tables, and as the analyser ran, the curves were followed by human operators who moved pointers geared to the mechanism, feeding in the varying values.

Output tables on the analyser produced graphs showing the solution to the equations set up on the differential analyser; in our simple example, the output would be a graph of distance travelled by the train versus time.　　　　　PJT

EMITRON CAMERA TUBE

HMV, Columbia and the Great Depression seem an unlikely combination to advance the story of television. But in the 1930s these protagonists joined the scattered assortment of inventors and scientists working world-wide to perfect the new medium, and developed a television camera which was to lead the field for years to come.

By 1930, HMV were well established as manufacturers of gramophones and records and had assembled a very strong team to research new methods of recording sound for domestic use and for the cinema. The interest in film led naturally to experiments with electronic pictures, and a team led by W F Tedham (1907–39) began to look into the possibilities of both mechanical and electronic television.

The Columbia Graphophone Company was HMV's biggest rival and much of their success at this time was due to the efforts of a young electronics genius, Alan Blumlein (1903–42), who, under the leadership of Isaac Shoenberg (1880–1963), had invented a system for recording gramophone records. This system, which later allowed the production of stereo records, prevented the company having to pay the crippling patent royalties faced by other companies.

In 1931, as the Great Depression began to bite, sales of both gramophones and records plunged. In order to survive, the two companies were forced to merge. The new company was called EMI, Electrical and Musical Industries, and the combined research team under Shoenberg was one of the most powerful collections of industrial research brainpower ever gathered together in Britain. As the new company took stock of the depressed marketplace, it became obvious that an outstanding new product was necessary to ensure survival. The research team was briefed to develop a viable television receiver.

In order to test the new electronic receivers, EMI also built a mechanical camera using technology already demonstrated by Baird (See LOGIE BAIRD'S TELEVISION APPARATUS). Two of the team, Tedham and McGee (1903–87), argued strongly that the company should also research the next logical step: the development of an all-electronic camera tube. Their request, however, was turned down and the pair were instructed not to waste their time on this project.

Tedham and McGee were so convinced that this tube could be made to work that they secretly constructed one in the autumn of 1932. They borrowed the equipment to test it, and after some minor adjustments the picture appeared as if by magic on the receiver. It surprised them both so much that they simply watched fascinated until the picture slowly faded as the tube failed, making no effort to record the achievement with a photograph. They knew now that a tube like this could be made to work, but as they had defied management instructions to build it, the now defunct tube was quietly put away in a cupboard. However, the two continued to argue for the go-ahead to develop it.

In 1933 Vladimir Zworykin (1889–1982), working for RCA in the United States, published an account of his electronic camera tube (the Iconoscope) which had given promising results. Despite the lack of detail in the description, it was obvious to Tedham and McGee that it was basically the same as their experimental tube of the previous year. Armed with this information they were at long last able to convince the management to authorize research into an electronic camera tube.

Despite a deteriorating financial situation, EMI expanded their research department and by the end of 1933, over sixty people were working on the development of television. McGee was able to demonstrate presentable pictures from a series of experimental tubes by the beginning of the following year and he instilled enough confidence for the company to cease development of mechanical cameras and concentrate on all-electronic television.

When the world's first regular high-definition television service started from Alexandra Palace in 1936, the three cameras in the EMI studio had at their heart the Emitron camera tube. The Emitron's task was to convert the light from the scene in the studio into a television signal. The way in which it did this was essentially the same as in the experimental tube made by Tedham and McGee four years previously. Even though it initially produced pictures which looked no better than its rival's, the Emitron had a number of very significant advantages, including the ability to produce good pictures in daylight (its rival, the Baird mechanical system, required powerful studio lighting). Although the reliability of the Emitron camera was not good by today's standards, it was infinitely better than Baird's system. Indeed, many of the early Emitron cameras were still in service nearly twenty years later.

The development of the Emitron enabled Britain to take the lead in television during this brief but critical pre-war period, and lay the foundation for the country's later pre-eminence in television broadcasting.

The original Emitron tube was essentially the same as the Iconoscope tube made by RCA in the United States, and in the fifty or so years since its development there have been several claims that the British tube was a straight copy of the US version. Although the extent to which the two companies collaborated is still unclear, the evidence indicates that the two tubes were developed completely independently. The technology of the day dictated that such a tube could be made in no other way, and this accounts for the similarity between the two.   JT

1985-5063

'MALLARD'

'Britain's Fastest Ever Express
Does 125 Miles an Hour'

Thus did the Daily Mirror announce the news of a high speed run by the streamlined steam locomotive *Mallard* on Sunday 3 July 1938. It is now accepted that *Mallard* briefly touched 126 mph and established the highest authenticated speed ever achieved by a steam locomotive. After running 1,426,621 miles in revenue earning service *Mallard* was withdrawn in April 1963.

If *Mallard* looks fast and exciting today it is a tribute to a design that astonished and excited observers when the first of the A4 class locomotives (of which *Mallard* was one) appeared in 1935. A total of thirty-five A4s were built between 1935 and 1938.

The streamlined A4s were a dramatic and effective response by Sir Nigel Gresley (1876–1941), Chief Engineer of the London & North Eastern Railway, to an increasing public demand for faster express passenger trains. All over the world between 1930 and the outbreak of war in 1939 there was a sudden and startling increase in railway speeds. In Germany, the diesel-powered *Fliegende Hamburger* streamlined train was running daily from Hamburg to Berlin at an average speed of 77.4 mph. In the United States larger diesel streamliners like the *Pioneer Zephyr* were providing high-speed luxury travel.

After observing German diesel traction and French high speed petrol railcars, Gresley put his faith in steam. He developed and refined his earlier express passenger designs of which the A1 class *Flying Scotsman* is probably the most famous. What distinguished the A4s was the streamlined casing which hid a conventional steam locomotive shape under a smooth carapace. Streamlining both reduced wind resistance and provided an eminently marketable shape with obvious style and panache. After exhaustive wind tunnel tests Gresley adopted a horizontal wedge shape for the streamlining similar to that used by Ettore Bugatti on his 1923 Grand Prix cars and subsequently in 1933 on a fleet of Bugatti railcars which Gresley

Newly in service,
Mallard climbs away from King's Cross Station, London,
with the Leeds express in 1938
(C C Herbert Collection, National Railway Museum)

had observed in France. The wedge helped with lifting smoke away from the driver's field of view, reduced air resistance to the locomotive itself and avoided disturbance to passing trains.

The exterior of the A4 locomotives was visibly different from the A3 design which preceded them but, to the engineer, the changes beneath the streamlining were even more significant. The boiler working pressure in the A3s was 220 pounds per square inch while in the A4s it was raised to 250 pounds per square inch. Very careful attention was paid to the size and shape of the steam and exhaust passages and to their internal streamlining. Gresley reduced the cylinder diameter on the A4s by half an inch to eighteen and a half inches and increased the piston-valve diameter to nine inches to give the maximum ease of steam flow between valves and cylinders. Thus, within and without, the locomotives were built for speed. The first of the class, *Silver Link*, ran some 10,000 miles in the first three weeks of service of which 7,000 were at an average speed of more than 70 mph.

Mallard left Doncaster works where it was built in March 1938. The locomotive was different from its

predecessors in having a double chimney and blast pipe which improved the exhaust of steam from the cylinders. Proof of the effectiveness of the system and of the quality of engineering that created *Mallard* was the 126 mph record achieved only a few months after the engine first ran.

It is the world speed record which gives *Mallard* its special distinction. In the hands of driver Joe Duddington and fireman Tommy Bray, with inspector Sid Jenkins riding on the footplate, *Mallard* was ostensibly taking part in one of a series of high speed braking tests. For the test on 3 July 1938 Gresley had decided, in great secrecy, to make an attempt to beat the speed record of 114 mph then held by the London, Midland & Scottish Railway. With a train of seven vehicles including a dynamometer car to record and measure the test, *Mallard* ran from King's Cross to just north of Grantham where it was reversed on a triangle of track. The speed record was achieved on the return journey, running down Stoke Bank between Grantham and Peterborough.

The calculations from the dynamometer record established, as the daily papers triumphantly claimed next day, 125 mph over a distance of a quarter of a mile. Although not mentioned in the press at a time when Nazi Germany was already threatening world peace, the Deutsches Reichsbahn record of 124.5 mph was just bettered. Subsequently a more detailed study of the *Mallard* dynamometer roll was able to show a brief few seconds at 126 mph.

. . . Once over the top I gave Mallard her head and she just jumped to it like a live thing. After three miles the speedometer in my cab showed 107 miles an hour, then 108, 109, 110 . . . Go on old girl, I thought, we can do better than this. So I nursed her and shot through Little Bytham at 123 and in the next one and a quarter miles the needle crept up further – 123½ – 124 – 125 – and then for a quarter of a mile . . . 126 miles an hour . . .

Driver J Duddington, speaking on the
BBC Home Service, April 1944
RSB

RANDALL AND BOOT'S CAVITY MAGNETRON

In 1940, the cavity magnetron was at the top of the secret list. Its military possibilities were such that there were grave doubts about the wisdom of telling even this country's closest allies about it. It was the key to smaller, more agile radar sets with vastly improved detection capability. On 21 February 1940, a cavity magnetron developed by J T Randall (1905–84, knighted in 1962) and H A H Boot (1917–83) worked for the first time in a laboratory at the Physics Department of Birmingham University. It was the culmination of an urgent search for a high-power source of ultra-short radio waves – those with wavelengths measured in centimetres.

In order to advance radar technology, the far-sighted Admiralty had, in 1939, placed a contract with Birmingham University for developing centimetre-wave generators; the task was overseen by the head of the Physics Department, the Australian Professor Mark Oliphant. He provided resourceful leadership and cushioned his scientists from the unfamiliar problems of military security and wartime stringencies, providing them with an environment where creativity could flourish.

On one occasion, they had failed to acquire a block of copper from their suppliers; being needed for war work, it was in short supply. Oliphant himself drove to the warehouse. When asked for a code which would release the material for the university's needs, he was temporarily lost for words. He read aloud his car registration number, the clerk wrote it down and the copper materialized. Some years after the war, Randall and Boot recalled their February 1940 triumph and modestly identified it as the one day when the other items of laboratory hardware connected to the cavity magnetron all happened to work at once.

Radar development in England had begun in earnest in 1935 (See ORIGINAL RADAR RECEIVER). The Chain-Home system devised by Watson-Watt's team operated at wavelengths of several metres. A basic rule of radio engineering states that an aerial system

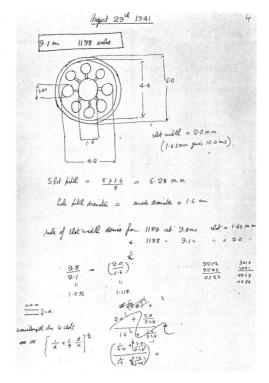

Laboratory notes on the magnetron from August 1941 (Museum of Science and Industry, Birmingham)

must, in order to form a directional beam of radio waves, have dimensions which are several times greater than the wavelength used. Consequently, the aerials used for the fixed Chain-Home radar stations measured over 100 metres in height and width. Clearly, more easily steerable, and mobile, radars had to have smaller aerials; equally, if accuracy and definition were not to be lost, shorter wavelengths had to be used. Radar engineers (including German ones) knew that centimetric (rather than metric) wavelengths would be best.

A fundamental problem with valves operating at short wavelengths is that the time taken for electrons to travel between the internal parts, the electrodes, is significant compared to the rate at which the radio signals are fluctuating in the external circuits to which they are connected. The cavity magnetron was developed by Randall and Boot to bypass this problem and reliably generate short wavelengths. Mathematical analysis of the electron paths in the cavity magnetron was one of the more complex and important tasks undertaken by D R Hartree's group, using the Manchester differential analyser (See MANCHESTER DIFFERENTIAL ANALYSER).

At first, it was difficult to measure the power and the wavelength of the radio waves generated by the cavity magnetrons. Filaments of light bulbs could be blown up and cigarettes could be lit. The consensus was that power of a few hundred watts, at least, was available. A few days later, it was established that the wavelength was 9.8 centimetres; they had aimed at ten. Bolder individuals in the laboratory found that the output of the cavity magnetron warmed up their fingers, presaging an important peacetime role.

Half a century after those days of swashbuckling invention, the cavity magnetron finds itself in more humble circumstances. One is at the heart of every microwave oven, producing the radio waves which heat the food. Despite scare stories about harmful radiation and poisoning from underheated food, microwave cookers are very much a part of domestic life in the closing years of the twentieth century.

The cavity magnetron transformed allied radar techniques. By 1941, ships were being fitted with centimetric radars which could often detect submarine periscopes. Soon afterwards, aircraft were equipped with cavity magnetron-based radars which improved the detection of other aircraft and provided ground-mapping systems for use at night and in cloud. 'Blind-fire' gunnery also benefited from centimetric radar. While the wartime application was the inventors' immediate reward, they could hardly have dreamt of the widespread peacetime applications of radar which followed after 1945.   ED

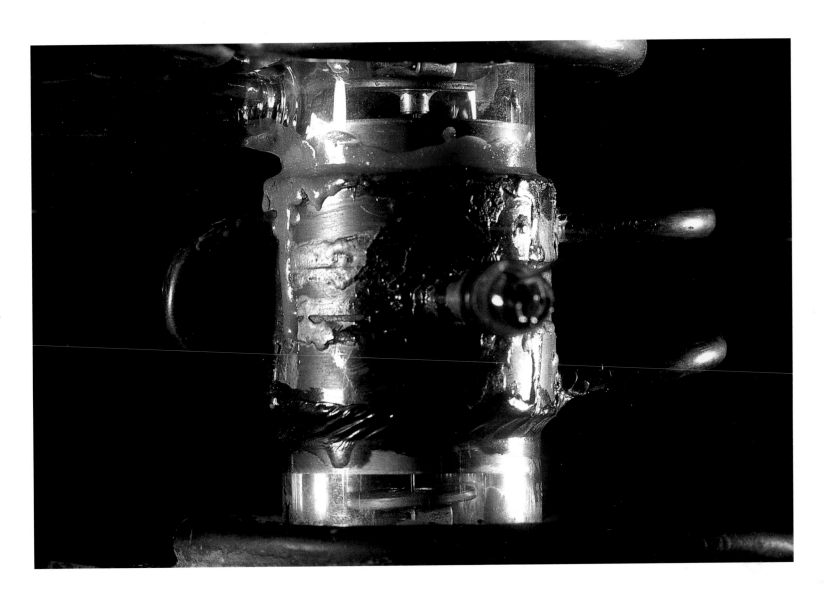

THE ROLLS-ROYCE MERLIN

The Merlin is uniquely famous. Perhaps the only aero engine to be widely known by name, it represented the contemporary pinnacle of engineering design and production skill. During the Battle of Britain the Merlin powered both types of defending fighters, the Hurricanes and Spitfires. In 1940 few were aware that the Merlin came from a long line of Rolls-Royce engines bearing names of birds of prey. Then the resonances were Arthurian, speaking of magic and power – an engineering magic which only Rolls-Royce could create.

Rolls-Royce had been drawn into the aero engine business at the government's request during the First World War. At first Britain had almost no aero engine industry and aircraft were equipped with imported French engines or with engines licence-built from French designs. Rolls-Royce undertook the manufacture of some of these types but regarded them as defective. Henry Royce therefore began work on an engine of his own that came to be known as the Eagle.

This was a 200 horsepower water-cooled engine with twelve cylinders in 'V' formation. Shortly after the war the dependability of these Rolls-Royce engines was demonstrated in a dramatic way by the first crossing of the Atlantic (*See* ALCOCK AND BROWN'S VICKERS VIMY) in an aircraft powered by two Eagles.

The experience of designing the Eagle passed into the ethos of the company. The directors even compiled the letters and memoranda from Royce concerning development of the engine into a book, *The First Aero Engine made by Rolls-Royce Limited*, 'as an example to all grades of Rolls-Royce Engineers, present and future.'

In the inter-war period the Air Ministry sustained three aero engine firms, sometimes referred to as 'The Family', by a considered rationing of contracts. Bristol manufactured high-powered air-cooled radial engines, favoured for the emerging airlines; Arm-strong-Siddeley also manufactured radials, though of less developed design, which were used in RAF training aircraft and in many military aircraft sold for export; Rolls-Royce built water-cooled in-line engines, chosen by aircraft designers for the clean aerodynamic nose they allowed on high-speed fighters.

By the late 1920s the Eagle had been replaced by the Kestrel – a 21-litre V-12 unit giving nearly 500 horsepower – which was fitted into the most modern fighters flown by the RAF. Curiously it was a Kestrel which was used for the first flights of both the prototype Messerschmitt 109 and the Junkers Ju 87 (Stuka) in 1935, before appropriate German engines were ready.

In this period Rolls-Royce also produced the 2,000 horsepower 'R' racing engines which achieved notable success in the Schneider Trophy races (*See* SUPERMARINE S.6B FLOATPLANE).

The Merlin, fruit of all this piston engine experience, was drawn up first in 1932. It was intended to develop 750 horsepower with the possibility of development to give 1,000 horsepower. Again, it used

Wartime installation of a Merlin in a Spitfire IV at the Vickers-Armstrong Castle Bromwich plant, September 1942

the V-12 layout and had a capacity of twenty-seven litres. It first flew in 1935. Later that year a Merlin powered the prototype Hawker Hurricane for its first flight and in March 1936 powered the first Spitfire. Throughout the ensuing war the Merlin was to undergo extraordinary development and almost every night throughout the war, the 'Derby Hum' – the drone of engines on the test beds – lay over the city. Power more than doubled over five years, mainly through ever-improving supercharger design, reaching well over 2,000 horsepower. The Merlin was also the subject of a major industrial effort with over 150,000 examples being produced by Rolls-Royce at Derby, Crewe and Glasgow, by the Ford Motor Company in Manchester and by Packard in the United States.

But the Merlin will always be best remembered for its role in the Battle of Britain, for this was a contest between engines as much as between aircraft. No other engine made anywhere in the world at the time could have given the Spitfires and Hurricanes the power and stamina to meet the Daimler–Benz powered Messerschmitts.

Both at the time and looking back over fifty years this battle is seen as one of the pivotal military contests in human history. It was the first check inflicted on German arms during the Second World War and as a result the United Kingdom survived to be the base for the Anglo-American assault on the Second Front. Modern Europe was born in 1940 over the fields of Kent. After the war, Lord Tedder, Marshal of the Royal Air Force, who had been in charge of the development of aircraft and engines during the period of the battle, attributed the British victory to three predominant factors: the skill and bravery of the pilots, 100-octane fuel, and the Rolls-Royce Merlin engine. There can hardly be a more powerful testimony to the social and human impact of engineering.   ALN

GLOSTER-WHITTLE E.28/39 JET AIRCRAFT

On 8 April 1941 the first British jet aircraft lifted off briefly while on taxiing trials and flew straight and level for about 200 yards. To the surprise of the Power Jets engineers who had built the Whittle W.1 power unit the flying men present seemed greatly relieved. Jet propulsion was so new that pilots had privately wondered whether the high-speed jet exhaust would make the aircraft squirm around uncontrollably 'like a dropped garden hose'.

After the trials the aircraft was taken back, not to the Gloster works, for fear of bombing, but to the premises of a provincial motor dealer, Crabtree's Garage, in Cheltenham. The uniqueness of the occasion was compounded when the pilot filled in the Test Flight Report and noted under the entry for type of propeller: 'no airscrew necessary with this method of propulsion'.

The first proper flight took place some weeks later, on 15 May, at Cranwell in Lincolnshire. The aircraft flew for some seventeen minutes and the most notable thing about this and the ensuing flights was the almost complete absence of trouble from the new engine. These flights gave a powerful boost to the view that the Whittle gas turbine programme was a 'potential war winner'.

Frank Whittle (b. 1907) was a cadet at the Royal Air Force College, Cranwell, when the idea for jet propulsion first came to him while writing a thesis on 'Future Developments in Aircraft Design' as part of his class work. Although he also reviewed the possible use of the gas turbine in the thesis he had not put these two ideas together and was thinking in terms of a jet produced by a piston engine buried within a hollow fuselage and driving a fan. He concluded that this would not offer any advantage over a conventional propeller but in the following year, 1929, he realized that a gas turbine could be used to generate a propulsive jet.

Whittle communicated his ideas to the Air Ministry but the feeling at the time was that materials were not available to stand the high temperatures in the turbine and that the development would be exceedingly difficult and expensive. Whittle was posted to test-flying duties at Felixstowe where he made numerous experimental catapult take-offs, and was commended as 'a very keen young officer and a useful test pilot'. In this period he continued to think about the jet engine and tried to interest various companies.

An outstanding performance on the Officers' Engineering Course persuaded the Air Ministry to send Whittle to Cambridge where he studied Mechanical Sciences. While at Cambridge he allowed his patents on the engine to lapse, but in 1935 he received a letter from a former RAF colleague, Rolf (later Sir) Dudley-Williams, c/o General Enterprises Ltd, Callard House, Regent Street. It said, 'This is just a hurried note to tell you that I have just met a man who is a bit of a big noise in an engineering concern and to whom I mentioned your invention of an aeroplane, *sans* propeller as it were and who is very interested. ... Do give this your earnest consideration and even if you can't do anything about the above you might have something else that is good.'

General Enterprises, the unlikely springboard for the British turbojet revolution, was a company mak-

ing and supplying cigarette vending machines which Williams ran with another ex-RAF man, J C B Tinling. It was the latter's father, a consulting engineer, who encouraged them to 'get into the aircraft business' due to the impending war.

The enthusiasm of Williams and Tinling helped bring about the formation of Power Jets Ltd to develop Whittle's ideas. Backing was found from a firm of City bankers and the Air Ministry agreed to let Whittle work as chief engineer for the company. The British Thomson-Houston Company, manufacturers of steam turbines, were subcontracted to build much of the engine and initially the tests took place in the BTH plant at Rugby.

By April 1937 the first ground-test engine, known as the WU (Whittle Unit), was running, although the problem of obtaining controlled stable combustion led to it being redesigned twice. By summer 1939 the engine was approaching its designed thrust. At this stage the Air Ministry became enthusiastic about the work of Whittle's team and design work for the Gloster E.28/39 was put in hand.

In spite of the success of this aircraft and its power unit the subsequent wartime development of the gas turbine was far from smooth. The decision was taken to put the engine into production but the design was altered to give more thrust and this modified version contained unproven elements. Furthermore the relationship was poor between the Whittle team at Power Jets and the Rover Car Company, which had been selected to manufacture the engine. Matters only improved when Rolls-Royce took over the project and the Gloster Meteor, the first British jet fighter, entered service in July 1944 in time to be used in action against the V1 flying bombs.

In 1944 Power Jets was nationalized, eventually making up part of the National Gas Turbine Establishment (NGTE) at Farnborough. The tiny company had just nine years of life but it had provoked an enormous change in aviation.   ALN

Above: The Gloster-Whittle E.28/39 at Farnborough in 1944.
Opposite: The Whittle W.1 engine used for the first flights

1946-110, 1946-26

THE V2 ROCKET

On 20 July 1969, Neil Armstrong and Buzz Aldrin became the first humans to set foot on the moon when they landed in the Sea of Tranquillity. They were launched on their voyage of exploration by the massive Saturn V rocket. Weighing 3,000 tons at launch, this vehicle had been developed in the United States by a team of engineers led by Wernher von Braun (1912–77). Some thirty years earlier, in Germany, von Braun and many of his Saturn V team had been working on a completely different rocket which would turn out to be the world's first long-range ballistic missile, the world's first space rocket and the forerunner of all modern space rockets. This was the German A-4 long range bombardment missile, later to be dubbed the V2 or Vengeance Weapon 2 by the Nazi propaganda machine.

Interest in rocketry had surged in the late 1920s in Europe and the USA, and many amateur rocket societies sprang up. Little official interest was shown in their activities, and their members were seen as impractical (and occasionally dangerous) enthusiasts. However, Germany's armed forces had been limited by the 1919 Versailles Treaty and were looking for ways to bypass these restrictions. The German Army Ordnance Department had appointed an engineer, Walter Dornberger, to investigate the rocket as a possible weapon of war, and in 1932, he contacted the German Rocket Society, VfR (Verein für Raumschiffahrt), who were then experimenting with rockets.

Although unimpressed by their technical work, one of the young members, Wernher von Braun, attracted his attention and in November that year von Braun joined Dornberger at the Army rocket test site at Kummersdorf, twenty-five kilometres from Berlin. He began work on a series of experimental liquid-fuelled rockets that would lead in just ten years from the early A-1 and A-2 rockets with thrusts of 300 kilograms and capable of heights of a few kilometres, to the A-4 rocket with twenty-five tonnes of thrust which could fly to a height of over 100 kilometres. The later rockets were far too large to be

A V2 is launched under British supervision
during 'Operation Backfire', Cuxhaven, Germany, 1945

developed at Kummersdorf and the establishment was moved to a secret site at Peenemunde on Germany's Baltic coast. The move was completed in 1940 and, on 3 October 1942, the first successful A-4 launch took place. This rocket reached an altitude of ninety-six kilometres and splashed down on target 192 kilometres away in the Baltic.

The dimensions of the A-4 had been fixed by the size of the railway tunnels in Germany, and Dornberger decided, rather arbitrarily, that it should have twice the range and more than ten times the payload of the German long-range Paris Gun of the First World War. The V2 missile was 14.3 metres tall, made of mild steel and weighed 12.8 tonnes at launch. Powered by a twenty-five tonne thrust liquid oxygen/alcohol motor it could carry a one-tonne payload over a range of about 300 kilometres. Its flight was controlled by four moveable carbon vanes in the rocket's exhaust, and 4.5 seconds after launch the guidance system began to tilt the rocket on to its pre-programmed flight path. After about sixty seconds of powered flight, when the rocket had

gained the velocity necessary to reach the target, some 5,500 kilometres per hour, the motor cut off and the V2 then continued flying on a free-fall ballistic path to its target.

As the first space rocket, capable of flying more than 100 kilometres up into the Earth's atmosphere, the V2 was a milestone on the path to the moon. But its effectiveness in World War II as a long-range missile is not so easily assessed. The V2 offensive lasted only from September 1944 to March 1945. About 6,250 V2s were produced and of these 1,115 reached Britain, half of them hitting London and causing 2,700 deaths and 6,500 serious casualties. A further 1,775 hit Continental targets, such as Antwerp, and 2,100 were in field storage at the end of the war. The remainder were either launch failures or test rounds.

Most V2s were made at the Mittelwerk, an underground factory in Central Germany where, at its peak, nearly 700 every month rolled off the production line. Forced labour from the nearby Dora work camp was employed at Mittelwerk and an estimated 20,000 inmates died whilst working there. The V2 came too late to have any effect on the outcome of the war, but its potential as a potent new weapon was widely recognized by the victors.

An unseemly squabble occurred at the end of the war to secure the secrets of German rocketry. The best engineering talent, including von Braun and Dornberger, with enough material to make about 100 V2s, went to the USA. Peenemunde and Mittelwerk were in the Soviet zone of occupation, and some important engineers and a supply of V2s went their way. Both countries eventually developed the V2 into the intercontinental ballistic missiles and the satellite launch vehicles we know today.

Britain gained experience of the V2 from the launching rather than the receiving end in October 1945 when, with the help of German technicians, three V2s were fired in 'Operation Backfire' from Cuxhaven. The V2 in the Science Museum was one of the unused Backfire rockets.   EJSB

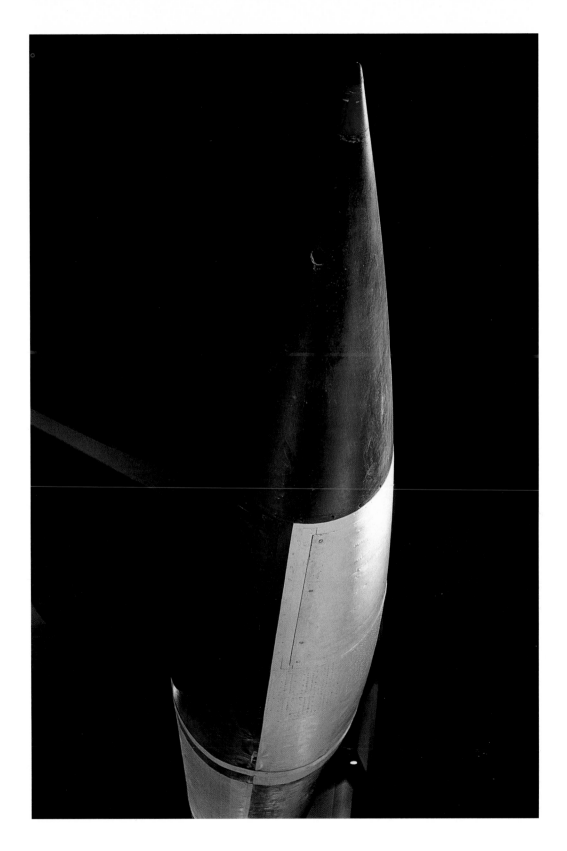

THE FIRST MARINE GAS TURBINE

The advent of steam propulsion at sea early in the nineteenth century and the subsequent growth of scientific tank-testing of hull shapes based on mathematical analysis confirmed one of the laws of physics which governs a ship's behaviour in water; shipbuilders had suspected it for a long time. What John Scott Russell, William Froude (*See* SWAN AND RAVEN) and their contemporaries were able to investigate, using improved experimental methods, was the relationship between increases in the power available to drive a ship and increases in her speed. Anyone who has rowed a dinghy or paddled a canoe will testify that a doubling of muscular effort does not produce a doubling in speed. In fact, it can be broadly stated that, for most vessels in water, power varies as the cube of speed. A fifty-horsepower engine which propels a fishing boat at a maximum speed of eight knots would have to be replaced by one of 169 horsepower to achieve a fifty per cent increase in speed to twelve knots.

Warship designers had long been preoccupied with this fact, since victory in a naval engagement will often go to the force that can outpace its enemy, manoeuvre itself into the most favourable attacking position and then retire quickly before the enemy has time to retaliate. But experience shows that a warship typically spends only ten per cent of its operational time at full speed; for the rest of the time while it is on passage from one area to another or carrying out a patrol, the output required of its main engines will be very much less. To meet this combination of requirements, naval engineers have tended to rely for 'cruising' power upon well proven engine designs known to be both reliable and economical in fuel consumption, supplemented where necessary by high-performance machinery whose high fuel consumption is acceptable if it can provide the extra speed which is crucially important in combat.

An early and innovative response to this requirement was the marine steam turbine first demonstrated at sea by Charles Parsons in 1894 (*See* PARSONS' MARINE STEAM TURBINE). Parsons' brilliantly simple invention rapidly found favour with all the major navies of the world, and derivatives of it continue to propel nuclear submarines a century later. Improvements in boiler performance early in the present century led to substantially increased power outputs from marine steam turbines, and these in turn prompted improvements in hull design, to enable high-speed combat vessels to remain at sea in weather conditions that were less than ideal.

Experimental work during the 1930s had established the very favourable power:weight ratio offered by the gas turbine. As the name implies, this later variant uses the energy of gas, which is produced by the combustion of fuel in an adjacent chamber. Since a heavy and voluminous steam boiler is not needed, the gas turbine offers superlative compactness, together with a very rapid availability of full power from a cold start. By the time Britain's first jet aircraft flew successfully (*See* GLOSTER-WHITTLE E.28/39 JET AIRCRAFT), the Admiralty were already beginning to consider applications of the same principle to warship propulson.

In August 1943 the Metropolitan-Vickers Electrical Company were commissioned to provide a variant of their gas turbine engine for an auxiliary power plant in a warship. With industrial production already flat out to meet the need for armaments and munitions, both manufacturing capacity and materials were in grievously short supply. To save time on unnecessary new development work, Metropolitan-Vickers used an existing F2/3 jet engine as a gas generator. Instead of an exhaust cone, the engine was fitted with a power turbine driven by the exhaust gases, coupled to a propeller shaft via a reduction gearbox. The engineers involved in this project were working at the limits of material science, and the high-performance alloys essential to this power plant were in very short supply. The war in Europe had already ended by the time all the technical and supply problems had been overcome.

Early in 1946 an existing motor gunboat was selected from the Coastal Forces fleet for the experimental engine, serially numbered G1. Of the existing three Packard petrol engines in the engine room, the centre engine was removed and replaced by the Metropolitan-Vickers gas turbine. Although rated at twice the power of the Packard engine it replaced, the gas turbine occupied no more space. A newly designed propeller was required to absorb the 2,500 horsepower on the centre shaft, and elaborate fire precautions were necessary to prevent ignition of any petrol fumes in the bilges when the gas turbine started up; this was achieved largely by airtight ducting.

By an accident of history, the trials of the Beryl engine in MGB 2009 took place in the Solent in July 1947, almost exactly fifty years after Charles Parsons had demonstrated his *Turbinia* at the Diamond Jubilee Review in the same waters. A comparison of the particulars of each vessel is interesting:

	1897 TURBINIA	1947 MGB 2009
Length overall:	100 ft	116 ft
Beam:	9 ft	20 ft
Displacement:	44 tons	100 tons
Power:	2,100 shaft horsepower	4,050 shaft horsepower

The speed achieved in each case was 34 knots.

In 1953 the Fast Patrol Boats *Bold Pioneer* and *Bold Pathfinder* emerged as the first warships to be built specifically for gas turbine propulsion, each employing two Metropolitan-Vickers G2 turbines. Today virtually all medium and large warships rely on some version of gas turbine to supplement their steam turbine, or more normally, diesel engines when full speed is required. Improvements in fuel efficiency have enhanced the reputation of the gas turbine as a compact power plant that can be delivering full power within less than a minute of being started. Such a capability will always be attractive both to those who design warships and to those who have to bring them home safely. JCR

PILOT ACE

The Pilot ACE (Automatic Computing Engine) was one of Britain's earliest stored program computers, and is the oldest complete general purpose electronic computer in Britain. It was designed and built between 1949–50 at the National Physical Laboratory (NPL) at Teddington in Middlesex.

The design of this computer stemmed from an earlier and larger computer designed by Alan Turing (1912–54) at the NPL between Autumn 1945 and Autumn 1947. Turing produced seven different logic designs for the ACE; ACE Marks I–V were designed and tested by Turing between Autumn 1945 and Spring 1946; and ACE Marks VI and VII were developed with help from James Wilkinson and Mike Woodger, two mathematicians at NPL. The original ACE models were essentially pen and paper exercises in mathematical logic. Though the machines were never built, Turing and his team would write and test programs on each design, using Turing's mathematical proofs and laboriously running each program through the intended logical operations bit by bit, and instruction by instruction.

Turing had gained previous experience of designing the logic for computing devices while working on the top-secret *Colossus* code-cracking machines during the Second World War at the government's Bletchley Park centre. He also had extensive contact with other computer pioneers working in England and the United States, including John von Neumann who pioneered the basic form of computer architecture that is still used in most computers today.

The National Physical Laboratory's original plan was to use outside contractors to construct the ACE. However, in 1947 an electronics group was set up there under H A Thomas, and it was decided that they would construct the ACE. Thomas was interested in industrial electronics, and did not want to share any of his precious resources with a mathematics-based project. The ensuing infighting caused

months of delays and problems, eventually leading to both Turing and the head of the maths team leaving in the autumn of 1947.

In 1947 the ACE team was expanded to include Donald Davis, an electronics expert, and Gerald Alway, a mathematician. Another new recruit was Harry Huskey who joined the team in January 1947 while waiting to take up a job at the University of Oklahoma. Huskey had worked on a technical description of the University of Pennsylvania's ENIAC (Electronic Numerical Integrator Analyzer and Computer), the world's first electronic digital computer. Huskey managed to persuade everyone, except Turing, that the only way forward would be to build a smaller version of the Mark V ACE design as a test bed for Turing's ideas.

Work on the test assembly started in the middle of 1947, but progress was slow as only a few members of the team had any electronic circuit design knowledge. By the time Huskey left in January 1948 most of the design work was complete, but no actual construction had started. In 1948 F M Colebrook replaced Thomas as the head of the electronics division. Colebrook was more sympathetic towards the ACE team and he suggested that they join the

The Pilot ACE computer at the time of its
first public demonstration in 1950

electronics team and work together. The new unified team worked well, and by late 1948 plans were made to produce a Pilot ACE – a pilot system which was part Mark V, and part test assembly.

Construction started in early 1949, the main chassis was completed by December, the control chassis by January, and by February there were enough working modules for the system to be capable of running some limited programs. The first full run was on 10 May 1950, the program switched the front panel lights on and off in succession. The simple test program consisted of a few instructions, which were entered by hand on the ACE's thirty-two front panel switches, as the punched card input/output unit had not been fully implemented.

A three-day press demonstration was held in late November 1950. The Pilot ACE worked perfectly throughout the demonstration running three demonstration programs with a bottle of beer for any member of the Press who could rival the machine in identifying a six-figure prime number. The ACE failed to work again for the next few weeks. Over the next six months the internal circuitry was changed and more delay lines were added. The next time it ran was on 26 June 1951 when at three pm it printed out the solution to seventeen simultaneous equations submitted earlier that morning.

The Pilot ACE was estimated to have cost £50,000 to design and build. However the cost was recouped many times over. In 1954 it earned over £24,000 on work which included computing bomb trajectories, calculating aircraft wing flutter, and testing crystallography theory.

The machine continued to work at the NPL until June 1956 when it was retired and presented to the Science Museum. The *Daily Mirror* reported this event under the heading 'The Government's first robot brain gets the sack'. The reported reasons were 'because faster robots are in use, and this one is so old and tired it's making mistakes'.   MA

NMR ELECTROMAGNET

As its cumbersome name suggests, nuclear magnetic resonance (NMR) is a complex technique. Its principle is straightforward however: it allows scientists to investigate the atoms inside a specimen. The process involves 'interrogating' the atoms by broadcasting radio waves of a precisely defined frequency from a transmitter close to the specimen and detecting and decoding the returned transmissions – the 'answers' – that these atoms give. Different groups of atoms in the specimen respond to particular frequencies in this radio transmission in a similar way to the well-known example of the wine glass responding to a particular note sung by a high-powered soprano. The response that the atoms give comes from their nuclei. These themselves can become radio transmitters, emitting a signal which can be picked up by a radio aerial mounted alongside. Precise measurement of the frequency and behaviour of this signal tells the experimenter what sort of atoms are present in the specimen, and in what numbers. More important, it can yield information about how those atoms are joined to others in a chemical compound, what other atoms are nearby, and even about the chemical reactions going on inside the specimen when the experiment is done.

For the technique to work, the atomic nuclei in the specimen must themselves be tiny magnets, as many nuclei are. These magnetic nuclei must also be lined up to point in particular directions, rather than lying in random orientations as they normally do. Alignment of the nuclei is done by exposing them to the field of a powerful electromagnet. Within the magnetic field produced by the electromagnet the magnetic nuclei line up rather like compass needles aligning themselves north-south under the influence of the Earth's magnetism.

The NMR apparatus in the Science Museum from which this magnet comes was built by Rex (later Sir Rex) Richards (b. 1922) and colleagues in an Oxford chemistry laboratory in the early 1950s. More than thirty kilometres of wire went into the hand-wound coils of the electromagnet. With its massive cast-iron yoke, it was the first magnet in Britain constructed specially for this new technique.

To investigate the shape and structure of molecules a liquid sample in a glass phial was placed in the narrow gap between the magnet's mirror-flat poles, together with the coils for transmitting and detecting the radio waves. Modern laboratory NMR machines look very different: smart anonymous boxes which can be bought off the shelf. Computer-aided to give instant results, and yielding much more detailed information than in the early days, they are among the most powerful tools available to researchers in chemistry and biochemistry.

Twenty years after Richards' pioneering work Peter Mansfield (b. 1933) and his partners at Nottingham University were building an electromagnet of different design. With water-cooled coils more than half a metre across (but no iron core), it was designed to apply the techniques of NMR to a very much larger specimen, the living human body.

Because water contains hydrogen atoms and hydrogen nuclei are magnetic, NMR can be used to detect water. And since the human body is two-thirds water, NMR can provide a non-invasive method of finding out about conditions inside the body. Different types of tissue contain different amounts of water, and diseased tissue differs from tissue that is healthy; NMR equipment can 'focus on' different areas within the body and give information about the type and condition of the tissue there.

Mansfield's first apparatus was the prototype for a novel technique of medical diagnosis, magnetic resonance imaging (MRI). It dates from the early 1970s. By 1980 his laboratory had developed an MRI system which could 'look at' different areas within the human skull, building up an image which showed a cross-sectional slice through the patient's head. Ten years later whole-body MRI machines had been installed in many hospitals around the world, often giving doctors a sharper and more informative picture of what is going on inside their patients than the X-ray CT scanner can (See THE FIRST BRAIN SCANNER), and seemingly without the health risk that comes with X-rays.

But for the country where Richards and Mansfield did their work, the story has a less than happy ending. In 1988 it was said that Britain had fewer MRI machines in its hospitals than any other country in the developed world. Not for the first time commercial exploitation of high-quality research seemed to have been left to overseas companies. British industry – and British patients – were losing out.

Increasingly for health managers world-wide the arrival on the market of the MRI machine revived a familiar dilemma. When funds are short, is it right to spend a large sum on a box of advanced technology wizardry – highly beneficial, but only to a comparatively few patients – or is it better to invest those resources in more down-to-earth forms of health care?   AWW

Opposite: NMR electromagnet built by
Rex Richards and colleagues

CRICK AND WATSON'S DNA MODEL

'We wish to suggest a structure for the salt of deoxyribose nucleic acid (DNA). This structure has novel features which are of considerable biological interest.' So began the short letter to *Nature* in 1953 announcing one of the most far-reaching discoveries of this century, the solving of the structure of DNA, the stuff of which genes are made.

Biotechnology and genetic engineering which have developed so dramatically in the latter part of the twentieth century owe their origins to this understanding of the structure of DNA and the ability to manipulate it. Disease resistant crops, specially designed drugs, scientific testing procedures, even treatments for hereditary illnesses have now become possible through these technologies. One of the most ambitious projects of the twentieth century is to map the entire human genome – to determine the genetic code of DNA in man.

In 1951 a young American research worker, James Watson (b. 1928), joined Francis Crick (b. 1916) at the molecular biology unit in Cambridge. Both were motivated by a growing conviction that DNA was the genetic material, and were determined to discover how its structure could enable it to pass on information from one generation to the next. It was already known that DNA was made up of long chains of alternating sugar and phosphate groups with further chemical groups (bases) attached to each sugar, but how were these long chains arranged in space?

The chief experimental evidence was emerging from the Medical Research Council unit at King's College London. Maurice Wilkins (b. 1916) and Rosalind Franklin (1920–58) were using the by then well-established technique of X-ray diffraction to study DNA's structure. When X-rays are shone on to a substance containing regular arrangements of atoms (such as is found in crystals) some rays are deflected and will form characteristic patterns of spots on a photographic plate. Analysis of such patterns had led to the solution of many crystalline structures over the previous three decades. The

fibrous structure of DNA also gave characteristic X-ray diffraction pictures indicating the presence of regular arrays of atoms.

Crick had previously shown mathematically that diffraction patterns of a helix (a corkscrew shape) took the form of a cross. The diffraction patterns obtained from DNA fibres by Rosalind Franklin showed just such a cross shape convincing Crick and Watson that DNA was helical. However it was not possible to obtain the complete structure directly from the X-ray data so far obtained, so Crick and Watson turned to model building in the hope that this intuitive approach might produce a structure compatible with both the X-ray data and the known chemical structure.

In their haste to be the first to solve the structure, they produced a model containing three helical molecules of DNA at the end of 1951. Wilkins and Franklin travelled from London to Cambridge to see it and, to Crick and Watson's chagrin, proceeded to demolish their arguments supporting this structure. During 1952 Linus Pauling (b. 1901) at Caltech in California also turned his attention to DNA. Having already discovered helical structures in proteins, he started building models of DNA, producing another triple helix structure towards the end of 1952. To Watson and Crick's enormous relief, however, this too was shown to be wrong. The race was still on. During 1952 Franklin produced her superb pictures of a second form of DNA. From these it could be deduced that the structure must contain two chains. It was Crick who then realized that Franklin's analysis indicated that the two chains ran in opposite directions.

The final breakthrough came when they considered some chemical evidence provided by Erwin Chargaff. Four different chemical groups, or bases,

could be attached to the sugar groups of the backbone; Chargaff had discovered that amounts of two bases, adenine and thymine, were always identical, he also found that the amounts of the two other bases, guanine and cytosine, were also equal. Watson realized that because of their shape and chemical nature these two pairs of bases could be arranged with very weak bonds holding them together (so-called hydrogen bonds). The shape of each pair was almost identical and would fit down the centre of a double helix formed by the two backbone chains.

Crick and Watson lost no time in building their model, using the metal plates now in the Science Museum to represent the four bases. This time there was no mistake and in the words of Max Perutz, the then director of the Laboratory of Molecular Biology, '1953 became the *annum mirabilis*. The Queen was crowned; Everest was climbed; DNA was solved'.

The elegant structure proposed offered an immediate explanation of how genetic information contained in DNA could be copied and passed from one generation to the next. The weak hydrogen bonds between each pair of bases which hold the two strands of the double helix together are easily broken. Each strand then acts as a template for building new molecules. The specific pairing of the bases ensures that each new chain is an exact replica of the original. Crick concluded his letter to *Nature* with the words: 'It has not escaped our notice that the specific pairing we have postulated immediately suggests a possible copying mechanism for the genetic material.'

In 1962 James Watson and Francis Crick received the Nobel Prize for Medicine together with Maurice Wilkins. Tragically Rosalind Franklin, whose beautiful X-ray diffraction pictures had provided key evidence, had died nearly five years earlier. By this time Crick had already started on his seminal work in unravelling the genetic code, enabling us to understand how DNA specifies the structure of the proteins it is programmed to make. AKN

Opposite: Reconstruction of
Crick and Watson's model made using
the original metal plates

FLYING BEDSTEAD TEST RIG

The rapid developments in aviation technology during the Second World War led to increases in the power, speed and weight of aircraft. Runways became longer, airfields became larger and hence more obvious targets for an enemy air strike. The development of guided missiles after the war made such permanent installations even more vulnerable.

These factors suggested to designers that an aircraft which could rise vertically, or from a very short take-off run, would have great advantages. Helicopters, of course, could do this, but in the late 1940s they were primitive devices: slow, unable to carry much payload and unsuitable for many military roles. But the war had forced the operational debut of the gas turbine and by the 1950s the new jet engines were showing markedly increased power outputs and improved reliability. The turbojet could produce a powerful airstream and it seemed that by directing the jet downwards a practical vertical take-off and landing (VTOL) aircraft might be achieved.

There was initially little official interest in VTOL aircraft in Britain but to Rolls-Royce, keen to exploit their new turbojets, the commercial attractions of VTOL were obvious. Their Chief Scientist, Dr Alan Arnold Griffith (1893–1963), was one of the pioneers of the gas turbine in Britain, which he had helped to advance while working at the Royal Aircraft Establishment, Farnborough. He had envisaged the use of jet thrust for VTOL as early as 1941.

Rolls-Royce decided to build two test rigs, which would demonstrate the principles of jet lift to doubters and also enable the development of the special control and stabilization systems thought necessary for the postulated family of new aeroplanes. Although officially called 'Thrust Measuring Rig' or TMR, the Heath Robinson appearance of the vehicle led to its being christened 'Flying Bedstead', and that name has been universally accepted ever since.

The prototype Flying Bedstead during flight testing at Hucknall in 1954

The rig was a tubular framework supporting two Rolls-Royce Nene engines of 5,000 pounds nominal thrust, each with their jet nozzles pointing downwards. A seat on top of the engines provided a lofty perch for the pilot. To provide control, tubular outriggers carried compressed air to small supplementary nozzles ('puffer jets') which could be operated by the pilot, an autostabilization system was also fitted because it was thought at the time that the pilot's reactions would be too slow to be effec-

tive. The total available thrust was 8,100 pounds, which provided little margin of safety in a machine weighing 7,500 pounds when carrying pilot and fuel.

The first TMR was rolled out at Rolls-Royce's airfield at Hucknall near Nottingham on 3 July 1953 and the first hovering trials followed a week later, with the rig suspended below a specially built gantry. Tethered tests continued over the next year until the first free flight was made by Rolls-Royce's Chief Test Pilot, R T Shepherd, on 3 August 1954. In the next few months the TMR made sixteen free flights, R A Harvey sharing the pilot's duties with Shepherd. The final flight at Hucknall was on 15 December 1954 after which the TMR was overhauled and transferred to the Royal Aircraft Establishment at Farnborough. In the meantime the second Bedstead, XK426, had been built. This began tethered hovers at Hucknall on 17 October 1955, and made its first free hover in November 1956.

Both TMRs were damaged in accidents in the autumn of 1957, the second of these killing a trainee pilot. Despite the accidents the Flying Bedstead programme demonstrated, in 380 tethered and 120 free flights, that jet lift could be used to support a fixed-wing aircraft during the take off and landing phases of flight, and that puffer jets provided adequate control during hovering. Rolls-Royce acquired the confidence to produce their first engine intended specifically for jet lift, the RB108, installed in Britain's first VTOL aeroplane, the Short SC1, which had one engine for propulsion and separate vertical units for lift. Subsequently the development of VTOL took a slightly different path with the Harrier in which swivelling nozzles were used to vector the thrust of a single main engine for lift and forward flight. However, it was the Flying Bedsteads which first pointed the way in Britain to successful jet-borne vertical flight. GM

THE CAESIUM ATOMIC CLOCK

To measure a quantity one needs a unit. Ideally this unit should be defined in terms of some naturally occurring phenomenon having universal rather than local significance. For the measurement of time such a natural unit exists in the length of the day but as this unit is too large for many purposes it has to be subdivided into hours, minutes and seconds. The length of the day depends on the rate of rotation of the Earth about its axis. Astronomical observations have shown that this rotation is not entirely uniform.

During the 1950s atomic clocks were produced whose timekeeping depended on a natural phenomenon, the vibration of caesium atoms. The frequency of these vibrations is so stable that in 1967 the General Conference of Weights and Measures redefined the second in terms of it, thus providing a constant unit of time for specific purposes. However our lives are still organized in terms of the day and for civil purposes atomic time is kept in step with the rotation of the Earth by adding or subtracting a leap second, in the same way that an extra day is introduced in a leap year to keep the calendar in step with the seasons.

An atomic clock consists essentially of a tube from which the air has been removed so that caesium atoms can pass along it. As they do so they are exposed to radio waves of a very high frequency (comparable to those used in radar). Each caesium atom acts as a small magnet and when the frequency of the radio wave corresponds to the natural frequency of vibration of the atom it flips over, reversing its polarity. A magnetic field deflects these atoms into a detector. By altering the frequency of the radio waves to maximize the number of atoms entering the detector it is possible to lock this frequency to that of the caesium atom. The frequency of the radio waves is derived, by multiplication, from a quartz oscillator which is thus controlled by the vibrations of the caesium atoms. The quartz oscillator can then

be used to indicate the time, as in the domestic quartz clock.

The idea of an atomic clock was put forward by Professor I I Rabi in his Richtmeyer Lecture to the American Physical Society in 1945 and most of the pioneering work was carried out in the USA. In 1948 Professor Polykarp Kusch (b. 1911), who worked with Professor Rabi at Columbia University, produced a design for a caesium atomic clock which was built at the National Bureau of Standards (NBS) in Washington. This instrument was operating by 1951 but it never realized the full potential of the design and was not sufficiently accurate to serve as a time standard. In 1953 Dr Louis Essen (b. 1908) of the National Physical Laboratory (NPL) in England returned from a visit to the NBS and other laboratories in the USA convinced that a successful caesium clock, based on Kusch's design, could be built at Teddington. Although the NPL lacked experience in atomic beam techniques Dr Essen had made important contributions to other fields (such as narrow-band width oscillators) which were crucial to the

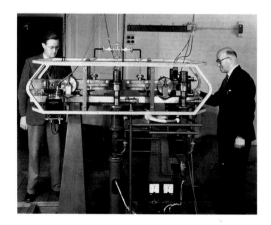

Mr J V L Parry (left) and Dr Louis Essen
with the atomic clock, 1955
(National Physical Laboratory)

development of the caesium clock. Assisted by J V L Parry he commenced work in the spring of 1953 and by June 1955 the clock was working reliably, with an accuracy which was equivalent to one second in 300 years. This was more accurate than astronomical observations which could therefore be supplanted as a standard. To establish a unit of time which was as close as possible to the existing unit the frequency of the caesium vibrations had to be measured in terms of the best value for the second which could be determined by astronomical observations. This work was carried out over the period 1955–8 in collaboration with Dr W Markowitz and Dr G Hall of the United States Naval Observatory in Washington DC. Their figure of 9,192,631,770 vibrations in one second was later used to define the second.

It is instructive to enquire why the NBS with their vast resources and expertise did not succeed in producing a reliable and accurate atomic clock. Paul Forman of the National Museum of American History, who has made an extensive study of this topic, suggests that this was partly due to the division of resources between two different atomic clocks (an ammonia clock and a caesium clock) which were being developed in parallel. Work on the caesium clock was also disrupted when the laboratory moved from Washington to Boulder in 1954.

Strictly speaking the NPL instrument was a time or frequency standard and not a clock since it did not run continuously or show elapsed time. However it is convenient to refer to it as a clock since it operated on the same principle as later atomic clocks which did run continuously and could show the time. Portable atomic clocks are now available commercially but they are still complex and expensive instruments. However, it has recently become possible to have atomic time in the home by means of clocks and watches which are controlled by radio time signals.   DV

THE WORLD'S FIRST HOVERCRAFT

During the nineteenth century it was recognized that a large proportion of the energy required to drive a ship through water was employed in overcoming the frictional efforts of the water on the surface of the hull. Numerous experiments were undertaken to reduce this friction either by pumping air over the surface of the hull to create air lubrication or by capturing an air bubble upon which the vessel rested, all to no avail.

In 1950 Mr C S Cockerell (b. 1910) (later Sir Christopher) left his employment as a research engineer with the Marconi company and acquired a small shipyard on the Norfolk Broads near Lowestoft. His background as an engineer led him to investigate the high frictional resistance of boats and after a number of experiments with air lubrication he came to the conclusion that the best method of tackling the problem was to 'float' the hull on a cushion of low-pressure air. Unlike previous attempts, which had always failed because of too much air leakage from the cushion when the craft heeled, Cockerell's method both contained the cushion and kept the craft stable. His breakthrough was to arrange a 'curtain' of high-pressure air ejected downwards from the rim of the craft.

To test this idea Cockerell built a working model in 1955 which used a model aircraft engine to provide both the supporting cushion of air and the containment curtain. It also provided forward thrust. The model worked well and Cockerell formed a company to develop the idea.

In 1956 the Research Branch of the Ministry of Supply began to take an interest in the project and placed a contract with Saunders Roe, the makers of many famous flying boats, to undertake a study of the concept in conjunction with Cockerell. This study was classified as 'Secret', and had as its main purpose the identification of the invention's military potential. The results indicated the viability of the concept although no sufficiently important military role was identified to warrant further military funding. In 1958 the project was declassified and the National Research and Development Corporation (NRDC) decided to finance further development work through a new company called Hovercraft Development Ltd. In the autumn of 1958 the NRDC placed a contract with Saunders Roe for the world's first full-sized hovercraft with the designation SR-N1 (Saunders Roe- Number 1).

The SR-N1 was conceived as an experimental craft from the beginning and upon its completion in June 1959 consisted of an oval platform thirty-one feet in length with a breadth of twenty-five feet, which hovered at a height of about one foot, so earning the nickname the 'flying saucer'. As in the model, the cushion, curtain and forward thrust were provided

Top: Testing the model shown opposite.
Above: The SR-N1 Hovercraft on trial, 1959

by a single piston engine, a 435 horsepower Alvis Leonides. The craft carried a pilot and an observer and demonstrated its ability to operate over both land and sea.

Although the SR-N1 crossed the English Channel from Calais to Dover in July 1959 it became clear as the experimental programme progressed that a hovering height of one foot restricted it to virtually smooth ground on land and two feet wave height at sea. In addition to these problems the shape of the platform led to difficulties in steering the craft. These directional problems were gradually solved by the provision of a separate jet engine to provide forward thrust and by the addition of a pointed bow and stern and an inflated keel. Height clearance was solved by developing the flexible skirt, a rubberized fabric hung from the platform rim through which the curtain jets were led. By means of the skirt the height of operation was raised from one foot to four feet.

The SR-N1 completed its experimental programme in early 1964 and from it a whole series of hovercraft were developed capable of operating over both land and sea. Perhaps the best known of these were the SR-N4 Mountbatten class cross-channel passenger ferries which entered service in 1968 and carried 250 passengers and thirty cars. These craft, in various extended versions, operated on this route until 1991.

Unfortunately the early promise held out by the hovercraft as both a mass transport vehicle and a new industry was not fulfilled in Britain due principally to the high costs of manufacture, maintenance and fuel, all of which were a feature of their aircraft ancestry. Although attempts were made in the 1980s to remedy these shortcomings, the hovercraft of the type represented by the SR-N1 has only been able to fill small specialized markets. Ironically, most development from the 1980s onwards has taken place abroad, mainly in the United States and the former Soviet Union, where its potential as a military assault vehicle has been recognized.   TW

AMPEX VR1000 VIDEO RECORDER

Magnetic tape recording, as distinct from previous systems using steel tape or wire (*See* POULSEN'S TELE-GRAPHONE), was developed during the Second World War in Germany. 'Magnetophon' recorders captured from the Germans, along with the secret recipes for the manufacture of the tape, formed the basis of all immediate post-war research.

The development at this time of instrumentation recorders capable of recording vast amounts of data encouraged the belief that video recording was a technical possibility. A number of establishments in Britain, North America, Germany, and Japan were engaged in research work towards this end.

By 1950, Ampex were well established as manufacturers of broadcast audio tape recorders in the United States and were developing a multi-track instrumentation recorder. At the start of 1952, a project to investigate video recording had been set up under Charles Ginsberg (b. 1920). He was soon joined by a young engineering student, Ray Dolby (b. 1933), whose name was later to become synonymous with noise reduction systems, and by the autumn the two of them were able to demonstrate almost recognizable pictures. The project was then abandoned for over a year (Dolby had been drafted into the army) but in late 1954 efforts were redoubled with a larger team who were able to solve the remaining problems within eighteen months.

One of the many technical obstacles to overcome was the problem of tape speed. In order to record the vast quantities of information in a television picture, the tape has to pass the recording head at a very high speed and it was on this point that most other designs foundered. In 1953, the Radio Corporation of America (RCA) demonstrated an experimental machine which ran at 360 inches per second and rushed one and a half miles of tape through the recorder to give just four minutes' playing time. The team at Ampex used a rotating head technique to solve the problem. Four heads mounted on a rotating wheel spinning at high speed scanned the slowly moving tape; although the tape

Early VR1000 installation at the BBC.
Note the racks of equipment surrounding the tape deck.
These house the record and replay electronics (BBC)

speed was only fifteen inches per second, the relative head-to-tape speed was 1,600 inches per second, giving one hour of recording on a ten-inch spool.

Ampex had kept this technological breakthrough a secret, so when they gave a surprise demonstration in April 1956 at the National Association of Broadcasters convention in Chicago, the industry was taken by storm and Ampex took orders for eighty machines within the first four days (an order book worth $4 million).

The first broadcast use of the machine was by CBS on 30 November 1956. By early 1958, a version of the machine was available in the United Kingdom and machines were purchased by Associated Rediffusion and Granada Television, with the BBC following suit the year after. The Ampex system was so successful that it rapidly killed off the opposition to the extent that several companies (notably RCA) signed licensing agreements to allow them to market machines using the same format. The Ampex two-inch format was to reign supreme in broadcast television for over twenty years and it fundamentally changed the way in which television companies operated.

In the mid-1950s, the vast majority of television programmes in the world were broadcast live, a method which involved complex extended rehearsals, shift working by production crews and presenters, and interludes to allow resetting of studios between programmes. Most importantly, programme choice and style was restricted to a format that would suit live transmission. From quiz shows to sophisticated drama, television was ephemeral.

Programme planners also had to cope with fixed events which had to be fitted into the schedule. So Saturday afternoon became the time for football and other sporting events; to time-shift an event like 'Match of the Day' was impossible without video recording. The headache of coping with a boxing match which might end half-way through the first round was equalled by the problem of what to do when a sporting fixture overran its time slot. Every television station had its stack of emergency caption cards. Phrases such as 'Normal service will be resumed as soon as possible' or 'We seem to have lost that programme... in the meantime here's some music' were both accepted and expected.

The move away from live television towards the present reliance on pre-recording took many years. This was partly due to the high cost of video recording technology in the 1960s and 1970s. As recently as 1971, BBC Leeds did not have their own video recorder but had to share one with BBC Manchester. Interviews for the evening regional news programme were often produced in the Leeds studio with the recorder on the end of a cable forty-five miles away, across the Pennines.

All this, of course, was only a precursor to the much greater television revolution following the introduction of the inexpensive domestic video recorder. The dramatic shift in viewing habits during the 1980s now means that the transmission time of a programme, or the clash of programmes on different channels, is no longer a problem for the viewer.

Thus this early Ampex video recorder is the milestone in the history of television which marks the beginning of the end of 'live' television, and the growth of new patterns of viewing.   JT

FERRANTI PEGASUS COMPUTER

The Ferranti Pegasus electronic computer is an example of an early computer that predates the use of transistors or modern integrated circuits. It is a tribute to the engineering design skills at Ferranti that this example, built in 1959, is still fully functional.

The Pegasus is most impressive to see and hear when running. Its valve circuitry consumes eighteen kilowatts of power provided by a noisy generator, while air conditioning units roar to keep it cool; and all this to achieve the computing power of a modern hand-held calculator. But it should be remembered that when it first appeared in 1956 it was capable of performing calculations infinitely faster than adding machines powered by human operators, the only alternative form of calculation generally available at that time.

The Pegasus computer used what was, for the time, a novel form of construction in which the circuitry was built from a number of modular packages. The package concept was first developed by Elliott Bros. at Boreham Wood and arose from a Navy requirement for an electronic gun-control computer. Previous computer designs involved assemblies of components wired together as one or two large units. However for shipborne operations it was not possible to carry spares of such large units, and so in the Elliott system the circuitry was built up from a large number of packages of which there were only a small number of different types. Each package within a type was interchangeable and hence only a small number of spares had to be carried since a faulty package could be replaced and then repaired elsewhere.

The package concept had many critics. Some said that the plugs and sockets needed at the rear of packages would be unreliable; others that the use of packages would restrict the logical design of a computer. All of these criticisms proved unfounded as Pegasus so successfully demonstrated.

The designer Christopher Strachey (1916–75) was

Pegasus computers being assembled at Ferranti's factory in Manchester, c 1958

at this time working at the National Research and Development Corporation (NRDC) and was an enthusiastic supporter of the package concept. Basing his design on experimental work he had done with Elliott Bros, he proposed the basic architecture for the Pegasus.

Strachey also proposed a new type of logic design, the 'order code' for the computer. This was a major advance in clarity and simplicity. The order code is the way in which a computer is instructed to carry out the very limited actions that it can perform. The computer uses binary information: ones or zeros. Humans cannot easily express themselves in this way; it is the order code which mediates between human intentions and the computer's internal processes.

In devising his logical structure for the machine Strachey was following the example of Maurice Wilkes (b. 1913), one of the leading British computer

pioneers, who designed and built the EDSAC computer at Cambridge in the late 1940s. Wilkes stressed the importance of the order code and placed great emphasis on ease of programming even at the expense of computing speed. As a result Pegasus was a machine that was easy to programme, logically very 'clean' and much loved by all who worked with it. Strachey's order code structure persisted into the late 1970s in the ICL1900 range of computers.

The first Pegasus machine was assembled in very elegant surroundings in Portland Place in London in 1956. It proved very reliable compared with other computers of the time and was soon providing the first bureaux service in Britain selling computer time to outside customers.

The Pegasus I evolved into the Pegasus II which, although of the same outward appearance, had improved circuitry, a larger drum store and more peripherals including a printer and magnetic tape units. In all, thirty-eight Pegasus I and IIs were sold, many going overseas. The Pegasus was one of the first computers to be produced on a production line set up in Ferranti's factory at West Gorton in Manchester.

The Science Museum's Pegasus, no. 18, was first sold to Scania in Sweden in 1959. In 1963 it came back to Ferranti whose computer division had by then become part of Standard Telephones and Cables Ltd. It then went to the Chemistry department of University College, London where Dr Judith Milledge used it until 1983 for the analysis of X-ray crystallography results. In 1983 it was acquired by the Science Museum though it was initially retained by International Computers Ltd at West Gorton where it was put on display to visitors. In 1989 it was restored to the Museum. This coincided with the formation of the Computer Conservation Society and former Pegasus maintenance engineers were contacted to set up a Pegasus Working Party. The machine was in full working order again by the summer of 1990.   AES

INTERNAL CARDIAC PACEMAKER

The development of the internal cardiac pacemaker is a tale of combined effort rather than individual endeavour. Several groups were working in the same area of clinical research at the time, all of whom sought to solve, in different ways, the problem of artificial pacing of the heart. The Science Museum's example is an early British contribution to the field.

Researchers had been active in this field for several years following the pioneering work of Paul Zoll in the USA. In 1952, he was the first to resuscitate a human heart using an electrical stimulus. In 1959 Elmquist and Senning, working in Stockholm, reported the first use of an implanted pacemaker in a human. Their unit was not completely internal, as it relied on an external power source to recharge the internal system. The first completely self-contained internal cardiac pacemaker was described in October 1960 by a team from Buffalo, New York State, comprising Chardack, Gage and Greatbatch. Other research groups were looking at a variety of solutions including using radio pulses or radioactive isotopes to produce a regular stimulus for the heart.

In London, a team based at St George's Hospital, including Drs Leatham and Portal and the surgeon Harold Siddons, were treating a number of patients with Stokes-Adams disease or heart block. A complete cycle of the contractions of the heart is made up of two separate stages. Firstly the auricles, the upper chambers of the heart, contract, pushing their contents down into the ventricles, the lower chambers. A second contraction then pushes the blood out of the ventricles into the arteries. In a normal heart the natural cardiac pacemaker generates an electrical pulse which initiates each cycle. This pulse is transmitted down through the wall dividing the two sides of the heart and on to the ventricles, leading to their contraction.

In a patient with heart block, the system conducting the pulse is impaired and this can lead to the heart temporarily stopping. When this occurs, the patient has a Stokes-Adams attack. Typically the person suddenly loses consciousness, falls to the ground and has a fit. The similarity of the symptoms meant that Stokes-Adams disease was occasionally confused with epilepsy.

In order to regulate the contractions of the heart artificially an electrode had to be inserted so that its tip was touching the inside of the right ventricle. This was achieved by cutting into the jugular vein and threading the end of the wire electrode down through the blood vessels and into the heart. A temporary external pacemaker was then connected to the other end of the wire, which remained outside the body, and to complete the circuit a second electrode was attached to the outside of the chest.

Once the patient was stabilized and the Stokes-Adams attacks had ceased, the patient was ready for a long-term pacemaker to be fitted. Initially the team at St George's experimented with small external pacemakers which could be carried around by the patient, but the problems with infection caused by keeping a permanent opening for the wires led them to try using internal models. The first design was made up of an implanted receiver and an external radio transmitter set into a belt worn round the waist. This was only partly successful, as it still required the patient to carry apparatus, and a powerful radio pulse was required before the internal receiver could pick it up. Geoffrey Davies, a senior cardiac technician at St George's Hospital, then designed the completely internal pacemaker which became the main device used at the hospital.

The unit, which was constructed in the physics department, was powered by mercury-zinc batteries which had an estimated life of five years. In clinical use it was intended that the whole unit would be changed every eighteen months. The pacemaker was implanted under the skin and two electrodes attached to the heart in a major operation where the chest wall was opened. Later it was found to be possible to set up the electrodes for long-term pacing by threading one through the jugular vein, as was done for temporary pacing. The pacemaker, and the second electrode, were then implanted under the skin and the electrode in the vein connected up in a much easier and safer procedure.

At first it was thought that the demand for cardiac pacemakers would be low, because of the relative rareness of the heart block, but later developments in the technology meant their use was possible in a variety of other cardiac conditions where the control of the heart's rhythm is affected. Recent estimates suggest that nearly one million people worldwide now have an implanted pacemaker.   JRC

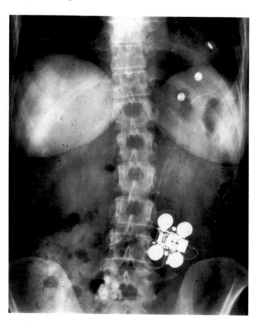

An X-ray showing a similar pacemaker implanted in the body (J G Davies)

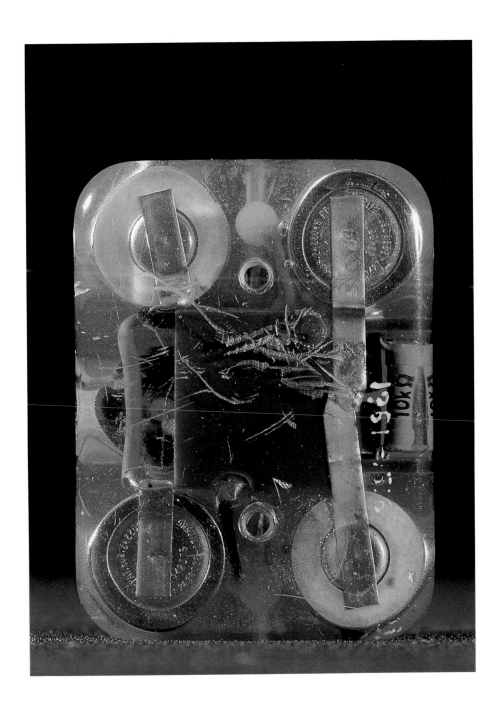

SCANNING ELECTRON MICROSCOPE

Intuitively, it might seem possible to see any object, however small, simply by enlarging its image in a microscope. But microscopes cannot give a clear image of objects which are much smaller than the wavelength of the illumination that is used to form the image. This means that we cannot see objects which are smaller than approximately one ten thousandth of a millimetre in a light microscope. To see smaller objects we must use a source of illumination which not only has a shorter wavelength but which can also be focused by lenses. Electrons satisfy both these requirements as the wavelength of a fast-moving electron is several orders of magnitude shorter than the wavelength of light and it can also be deflected by an electromagnet, reproducing the effect of a lens.

The first electron microscope was built by Max Knoll (1897–1969) and Ernst Ruska (1906–88) in 1931 and evolved from a research project on the high-voltage oscillograph at the Technische Hochschule in Berlin. This instrument operated in the same way as a light microscope: the electrons passed through the specimen and an enlarged image was formed by the electromagnetic lenses. The scanning electron microscope however builds up the image, bit by bit over a period of time, like the image on a television screen. This is achieved by scanning a fine beam of electrons, with a zigzag motion, across the surface of the specimen, and collecting the electrons knocked out of the surface of the specimen by the electron beam. These electrons are used to control the brightness of a spot of light – on a video screen – which moves in synchronism with the beam. The scanning electron microscope therefore has the advantage that it can examine the surface of specimens directly. The size of the electron beam determines the fineness of the detail and the magnification can be altered by varying the area of the specimen which is scanned: the smaller the area the higher the magnification.

Scanning electron microscopes operating on the

A scanning electron micrograph, showing the depth of field in such images (Philips)

principle described above were constructed by Manfred von Ardenne (b. 1906) and Max Knoll around 1933–4. In the following years von Ardenne attempted to see finer detail by using an electromagnetic lens to reduce the size of the electron beam, but the quality of the image was poor. In 1948 Charles Oatley felt that it was timely to reconsider the feasibility of the scanning electron microscope in the light of advances in electronics which had occurred during the war and shortly afterwards. He initiated a research project at the Engineering Department of the University of Cambridge and by 1951 his research student Dennis McMullan had produced a working instrument with a resolving power which exceeded that of the light microscope. This was the forerunner of the modern scanning electron microscope. This instrument was taken over by another research student, K C A Smith, who dramatically improved the efficiency of the electron collecting system.

The Canadian Pulp & Paper Research Institute was impressed by the results obtained with the improved microscope and they employed Dr Smith to construct a properly engineered version which worked successfully for many years in Canada. Despite the success of the Canadian instrument a further seven years was to elapse before the instrument was produced in quantity. It was felt at the time that there would not be a sufficiently large market for it since the conventional electron microscope, which by then was firmly established, could see finer detail and examine surfaces indirectly. Lower magnifications appeared to be adequately catered for by the ubiquitous light microscope.

Eventually the Cambridge Instrument Company agreed to manufacture the scanning electron microscope, influenced by the fact that some of the parts would be common to another instrument which they were already producing. One of Professor Oatley's research students, A D G Stewart, joined the company to work on the microscope, thus maintaining the link with the university. To test the market the company decided to produce a trial batch of five instruments in 1965. The Science Museum's example is the fourth instrument, which was delivered to the Central Electricity Research Laboratories.

The five were sold under the same trade name 'Stereoscan' which was chosen to emphasize the strikingly three-dimensional quality of the images. This was enhanced by the tremendous depth of field which was far greater than that of the light microscope at the same magnification. It was only when these instruments came into the hands of microscopists that this feature and others such as the ease of specimen preparation were fully appreciated. (It was found later that a high proportion of the work carried out with the instrument was done at magnifications within the range of the light microscope.) Sales boomed and similar instruments were soon produced by other companies. By 1985 annual production had risen to 1,000 instruments, in contrast with a market survey conducted in the USA in 1963 which predicted that ten instruments would saturate the market. Scanning electron microscopes now outnumber the non-scanning variety.   DV

ULTRASOUND SCANNER

In Britain it is now standard practice for pregnant women to have ultrasound scans. Yet the first experimental ultrasound machine for use in gynaecology and obstetrics was only made in 1956. In thirty-six years a novel experimental procedure has become part of the normal experience of pregnancy. The 'Diasonagraph Mk1' from the Institute of Obstetrics and Gynaecology in London played a key part in this revolution.

Ultrasound is sound higher in pitch than can be heard with the human ear. An ultrasound scanner uses crystals which naturally resonate at these high frequencies to transmit and pick up sound waves. In the first experiments in ultrasound, undertaken during the First World War, the French physicist Paul Langevin (1872–1946) aimed to develop a technique for detecting submerged German submarines. Work on the principle has continued since, with important developments occurring during the intensification of submarine warfare in the Second World War. From 1928 the principle was also applied to detecting flaws in metal components.

The first published experimental ultrasound examinations of women, undertaken by Professor Ian Donald (1910–87) and his colleagues in Glasgow in the late 1950s, used an industrial flaw detector to show that echoes from within the patient's abdomen could be used to measure the size of ovarian cysts and other tumours. This is known as a one-dimensional scan because it shows how far apart things are without producing an image. In two-dimensional scans actual images are produced by the technique of moving the ultrasonic probe in a series of sweeping movements during scanning. Both techniques are now used when you go for a 'scan'.

When ultrasonic waves are passed from a scanner into the body, they bounce back like echoes when they meet the interfaces between different kinds of tissue. The scanner picks up these reflected sound waves and displays them on a screen. Structures that are nearer the surface of the body send back echoes before those which are deeper, so what appears on the screen corresponds exactly to what is inside the body. Structures which are denser send back stronger echoes, which allows doctors to distinguish bones from muscle, for example.

Professor Stuart Campbell's (b. 1936) work using the machine now in the Science Museum, made in about 1967, established many of the procedures which have made ultrasound an indispensable part of routine antenatal care. In a period of heightened awareness of the dangers of X-rays, ultrasound was seen to offer a safe way of examining all pregnant women for multiple foetuses and malformations. For example, in 1969, quintuplets, later successfully delivered, were diagnosed at nine weeks using the two-dimensional scan. The first abortion of a malformed foetus after an ultrasound examination occurred after this machine disclosed a case of anencephaly (the partial or total absence of a brain)

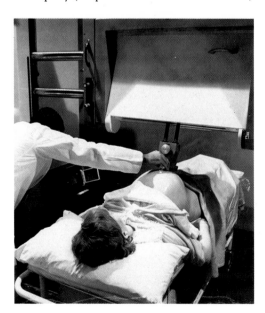

Professor Ian Donald supervising an ultrasound scan at the Queen Mother's Hospital, Glasgow, 1967, using a scanner similar to the one shown opposite (British Medical Ultrasound Society Historical Collection/Scottish Daily Express)

in 1972. Diagnosis of spina bifida followed in 1975.

Whilst checking for gross abnormalities and multiple pregnancies is now an important part of a scan, perhaps more significant in the majority of cases is the use of these machines to produce measurements of the foetus which allow comparison with established ideas of growth. During the late 1960s and the 1970s Campbell, using the Diasonagraph, developed a number of these techniques. In London he built on earlier work he had undertaken with the Glasgow team in which the foetal head was measured by first accurately establishing its position in the womb using a two-dimensional scan, then measuring it using the one-dimensional technique. By comparing the size of the head with established norms it was possible to judge whether the foetus was the right size for its age, or to estimate its age if the date of conception was unknown. Other techniques followed, including one which worked out the foetus' hourly urine production from the volume of the bladder. A low urine production rate was found to be linked with reduced foetal growth. It was also possible to 'weigh' the foetus whilst in the womb by measuring its abdominal circumference, once again relating this reading to normal growth rates to establish whether a foetus could be considered small for its age.

The rapid acceptance of ultrasound scanning, associated with other new obstetric technologies, notably foetal monitoring during childbirth, has not passed uncriticized. Some groups have argued that by inventing a technology which makes it possible to 'get to know' the foetus independently of the mother, doctors have bypassed the one person who might be expected to know most about the pregnancy. However, recent studies have indicated that having a scan helps some mothers to bond with their babies. It is also entirely possible that this machine, because of its role in establishing the core techniques of antenatal ultrasound, has played a part in helping patients to reflect on the part played by machines in their medical treatment. TMB

THE FIRST FLEXIBLE MANUFACTURING SYSTEM

Molins System 24 was a British manufacturing concept ahead of its time, which brought together the newly emerging technologies of the computer and the numerically controlled machine tool into an integrated system that could manufacture components with very little human intervention. The result was one of the biggest single developments in the history of production technology in the twentieth century.

Modern mass production techniques trace their ancestry back to the small workshops which appeared during the first half of the nineteenth century, and although the machine tools – the lathes, milling machines and so on – have become more sophisticated, they are essentially the same as those designed by Whitney, Whitworth and Maudslay. The Portsmouth block-making machinery (See PORTS-MOUTH BLOCK-MAKING MACHINERY) was the world's first mass production line (c 1803) and this basic concept remained largely unchanged until the middle of the twentieth century. The advent of numerical control in the early 1950s, whereby machines are controlled automatically by coded instructions, helped to improve the efficiency of machining, but did little to streamline the progress of the component through the manufacturing process. The introduction of the digital computer, in the late 1950s, was the integrating element that could replace human control of individual machine tools and enable the whole manufacturing process to be entirely automatic.

In 1965 a research and development engineer, D T N (Theo) Williamson (1923–92), working for the Molins Machine Company, tobacco machine manufacturers in Deptford, south-east London, first integrated these elements into a computer-controlled flow-production system, the System 24. Williamson had been one of the early pioneers of numerical control (NC) in the United Kingdom, and had published the first British paper on the subject in 1955, whilst working for the electronics firm Ferranti in Edinburgh (See FERRANTI PEGASUS COMPUTER).

The first prototype System 24 machine tool began to cut metal in 1964, and a further range of models were developed between 1966 and 1968. A series of patents were granted from about 1970. The concept of System 24, as Williamson envisaged, was that of a number of complementary machine tools ('units') under the control of a small digital computer which also controlled a conveyor system, so that workpieces could be automatically moved from unit to unit for different machining operations. The intention was for the system to run unattended throughout a sixteen-hour night shift. Maintenance operators were only required during the day to load and unload the system. The computer controlled all the handling equipment and the processes of workset-

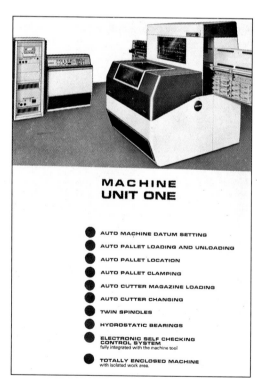

MACHINE UNIT ONE

- AUTO MACHINE DATUM SETTING
- AUTO PALLET LOADING AND UNLOADING
- AUTO PALLET LOCATION
- AUTO PALLET CLAMPING
- AUTO CUTTER MAGAZINE LOADING
- AUTO CUTTER CHANGING
- TWIN SPINDLES
- HYDROSTATIC BEARINGS
- ELECTRONIC SELF CHECKING CONTROL SYSTEM
  fully integrated with the machine tool
- TOTALLY ENCLOSED MACHINE
  with isolated work area.

A page from the Molins System 24 brochure, 1971, showing the Unit 1 machine (Molins plc)

ting, resetting, inspection, unloading, program tape selection, cutter changing, and loading and rescheduling in the event of machine failure. Each unit in the line had different functions, and held magazines of between eleven and twenty-eight tools, but taken as a whole the line was capable of complex metal-cutting operations, all linked by an accurate component transfer and location system. In addition the system was to use binary coding for identifying the pallets which carried the components from one location to another. The system was designed to manufacture parts from aluminium alloy and to achieve as high a productivity rate as possible.

Only one full prototype system was manufactured, for IBM, but components of System 24 were sold to a number of companies, including Texas Instruments and Rolls-Royce. British Aerospace (Wharton Division, Preston) had six Unit 1 machines up to the mid-1980s, arranged in a cell system, similar to the original Williamson concept. However little else of Williamson's original idea was effectively put into practice, as technological advances in the United States in cheap micro-electronics during the late 1960s and early 1970s, enabled companies like Cincinnati Milatron to develop simpler, less costly manufacturing systems.

In the end the overall concept proved too expensive for Molins to develop to its full potential and the company eventually dropped the scheme. However, the influence of System 24 quickly spread throughout the production industry. Remnants of Williamson's ideas can be seen today in advanced FMS (Flexible Manufacturing Systems) and in the field of CIM (Computer Integrated Manufacture). As long ago as 1972 Williamson wrote *The Anachronistic Factory*, in which he anticipated the advent of the automatic factory many years before it became a reality. Undoubtedly the developments in automated manufacturing which we now see throughout the world owe a great deal to the pioneering work and original thinking of Theo Williamson and his team at Molins.

JG

CONCORDE 002

The origins of Concorde can be traced back to the early 1950s. Jet engines were developing rapidly, while swept and delta (triangular) wing shapes were being explored for supersonic military aircraft. In March 1956 the potential of the new technology was demonstrated when the Fairey Delta 2, powered by a Rolls-Royce Avon turbojet, gained the world air speed record at the then astonishing speed of 1,132 mph (Mach 1.7).

The first jet-powered civil airliner, the de Havilland Comet, which entered service in 1952, brought a significant reduction to international journey times. The trend in civil aviation since the end of the First World War had been towards faster aircraft speeds and it seemed a rational step to consider supersonic travel. Design studies were initiated by the Royal Aircraft Establishment (RAE) at Farnborough.

In November 1956, the Supersonic Transport Aircraft Committee was formed and in 1959 recommended a major step forward: the construction of a fleet of long range inter-continental airliners that would fly nearly twice as high (50–60,000 feet) and over twice as fast as those in service at the time. The design speed was to be Mach 2.0, twice the speed of sound.

France had also been pursuing the idea of a supersonic airliner and in 1962 the British and French Governments signed a binding agreement to develop the aircraft together. The designers were faced with many new problems and the two prototypes, 001 (French) and 002 (British) were built to put theory into practice.

Above all else Concorde had to be a safe aircraft and, from a pilot's point of view, it had to handle in a conventional manner at all speeds; flying at the speed of sound (Mach 1.0) and at twice that speed, the equivalent of about twenty-three miles per minute, had to be no different to flying at subsonic speeds. Every possible problem was thought through and back-up systems provided for every service that might fail; for example, there are four different methods of lowering the undercarriage.

A wind tunnel model of Concorde on test
at the Royal Aircraft Establishment, Bedford, in 1962

A maximum speed of just over Mach 2.0 was chosen because aluminium alloys could be used. At this speed Concorde's skin temperature can reach 120 degrees centigrade due to air friction. To fly at higher speeds titanium and stainless steel would be needed, thus involving more costly research and development.

Concorde 002's complex and elegantly curved wing was designed to meet the requirements of low drag at high speed and high lift at low speed. Furthermore, it contains eighty per cent of the total fuel load of 105,000 litres.

Many innovative technical solutions had to be developed for the programme. Concorde's delta wing gave a pronounced nose-high attitude at low speed and to improve the pilot's view when landing and taking off, a mechanism was devised to drop the nose section forward of the cockpit. In its raised position the nose gave a clean aerodynamic shape at high speed. The engines required sophisticated control systems in intake and exhaust passages to 'tune' airflow for low speed or supersonic flight.

Concorde 002 made her maiden flight from Filton in Bristol to Fairford in Gloucestershire on 9 April 1969, piloted by Mr Brian Trubshaw. Over four hundred test flights were made. Mach 1.0 was achieved on 25 March 1970 and Mach 2.0 on 12 November 1970.

By 1972, sufficient confidence in 002's performance had been gained to undertake a demonstration tour. In June the aircraft visited the Middle East, the Far East and Australia flying 45,000 miles and visiting twelve countries in thirty days; it was a remarkable achievement and 002 created tremendous interest wherever she went.

However, serious concern about cost, commercial viability, sonic booms and engine noise was growing. It also became apparent that most national governments would not allow supersonic overflights of their territory. The dream of fleets of supersonic airlines encircling the earth and bringing continents and peoples closer together foundered on these objections and in the end a total of only twenty Concordes were built; of these, six were used for trials and fourteen entered airline service (seven with Air France and seven with British Airways).

To some, the development of Concorde was 'one of the worst investment decisions in the history of mankind' (David Edgerton). Others still view it as a pioneering European venture in aircraft building which advanced British industry to a new level of technological competence. In any event the development of the aircraft was a work of enormous technical virtuosity. Concordes 001 and 002 provided the step that led to Concorde airliners regularly crossing the Atlantic in three hours fifty minutes, far above the weather and at twice the speed of sound.   PC

ROLLS-ROYCE RB 211

When the jet engine was developed during the Second World War it was seen at the outset as a power plant for fast military aircraft. The exhaust of a simple jet engine consists of a very high-speed airflow and if this is used to power a correspondingly high-speed aircraft it can be efficient. However, when powering relatively low-speed aircraft, such as a conventional airliner, the propulsive efficiency is low and energy is wasted. The reason is that the exhaust leaves the jet engine with excess speed relative to the surrounding air and its energy is then dissipated in the atmosphere by turbulent mixing – the source of the intense noise associated with the older generation of jet airliners.

Contrary to expectations the jet engine was adapted for civil use in the years shortly after World War II. Initially its inefficiency for this use did not seem to matter since the new passenger jets like the de Havilland Comet and the Boeing 707 offered a great improvement in smoothness, internal noise levels and speed. However, as fuel costs rose and the public opposition to noise at airports grew, there was growing pressure on aero engine manufacturers to tackle these problems.

Fortunately the solution to both defects was the same, and had been proposed by, among others, the gas turbine pioneer A A Griffith (1893–1963) in the 1930s (See FLYING BEDSTEAD TEST RIG). The principle was to adapt the gas turbine to produce a high volume of relatively slow-moving air.

In such a 'high bypass ratio' design the core engine takes only a small proportion – perhaps a quarter – of the total air breathed by the engine and serves as a workhorse to drive the rest of the air around it with the multi-bladed ducted fan. The apparent drawback of such a scheme was the air resistance and weight of the large nacelles which would house the fan. The 'performance engineers' at Rolls-Royce who oversaw the installation design of engines in customer aircraft believed that these problems would cancel out the gains in propulsive efficiency.

As the originator of the RB 211 concept, Geoffrey

Vickers VC10 flying test bed.
The two port engines removed to fit
one RB 211 on test

Wilde (b. 1917) had done performance calculations which indicated to him that a very high bypass ratio would work and could not understand the opponents of this solution, recalling that 'after all, they could do the same sums as me'. However, Rolls-Royce only became committed to really high bypass ratios when it became clear that the American aircraft builders, and particularly Boeing, were convinced that it was the correct technical path to follow for the coming generation of wide-body ('Jumbo') airliners. The general architecture of the RB 211 began to take shape in 1967.

The development problems faced by Rolls-Royce were immense. The size of many components was more than twice anything they had built before, while to obtain the required efficiency from the core engine the operating temperatures had to be higher than in previous Rolls-Royce engines. In addition, the decision had been made to make the large fan out of Hyfil – a new carbon fibre composite material.

By 1970 it became clear that the programme was in trouble. The Hyfil fan blades, while immensely strong in tension, could withstand neither the mandatory bird-strike test nor rainwater erosion, necessitating a re-design in titanium. This was one of numerous problems which meant that development

proved more difficult and expensive than had been foreseen.

A crisis loomed: Stanley Hooker, the designer of the Merlin supercharger (See ROLLS-ROYCE MERLIN) during the Second World War (and subsequently Chief Engineer in the engine division of the Bristol Aeroplane Company) was brought in to take charge of the engineering effort. Rolls-Royce then reapplied itself to the task it has always done superbly well – painstaking development. However, the company had overspent and although the improved engine was almost ready Lord Cole, Chairman of Rolls-Royce, had to declare the company insolvent.

The company was rapidly acquired by the government which agreed to finance the day-to-day costs of development. Hooker recalled, 'We gradually got the great company not merely on the road again but really humming. Bankruptcy may temporarily shatter morale but it certainly concentrates the mind wonderfully'. Deliveries of the engine began in 1972.

The RB 211 has gone on to be a great success, often selected by airlines over its rivals from General Electric and Pratt & Whitney for a variety of aircraft, because of its good economy and durability.

The launch of the RB 211 was a unique event in aero engine history. There had been rapid evolution in the twenty-five years since the Second World War, with large leaps in performance in each new generation of engines. The big fan engines may prove to be the last to be developed from scratch as a leap into the unknown. Today, aero engine design is mature and performance gains are won from incremental improvements to a basic design – a less risky strategy.

The RB 211 has proved to be a crucial engine for Rolls-Royce. It has provided the basis for subsequent development of all their large engines. Without a fan engine suitable for aircraft such as the Boeing 747, the company could not have remained a major engine maker. The engine which bankrupted the most famous British engineering company also ensured its survival in the front rank of international competition.   ALN

THE FIRST BRAIN SCANNER

This brain scanner from Atkinson Morley's Hospital in Wimbledon was the first to be made and the experimental model with which the earliest trials on patients were undertaken. Techniques developed with this machine established computed tomography (CT scanning) as a key imaging technology, particularly as a tool for studying the brain. CT scanning is a method which produces an image of a 'slice' of the body, using a computer program to reconstruct a picture from data derived from a series of precision X-ray exposures.

From its formal announcement in 1972, the CT scanner was a runaway success and by 1977 there were 1,130 machines installed in countries across the world. A large part of this success can be traced to the fact that it provided diagnostic information that was not available by any established technique. A conventional X-ray picture is an image directly produced on a photographic film by X-rays which are absorbed to different extents according to the differing densities of tissues they pass through. That means that in an X-ray picture of parts of the body which contain many organs, images of all of them are laid on top of each other, making the images difficult for radiologists to interpret. Since the earliest days of radiology, techniques have been invented to overcome this and to show up one organ against another, for example by asking patients to swallow a barium meal to reveal the contours of the stomach or gut separately from the other organs of the abdomen.

Tomography, originally developed in the 1920s, used a mechanical method of taking pictures of particular layers of the inside of the body. It worked by moving the X-ray tube and plate in opposite directions whilst the X-ray was being taken, causing all the image layers but the one in the centre to become blurred. Most tomographic X-ray machines produced pictures of layers parallel to the body's length. A method for producing X-rays of 'slices' across the body was developed in the 1940s, but it was never really more than a scientific curiosity.

However, it was developments in computing, not in radiology, which led to a clinically useful form of tomography. The inventor of the scanner, Godfrey Hounsfield (b. 1919), had previously worked on computer associated projects at EMI, including the EMIDEC 1100, the first British transistorized business computer, and novel types of computer memory devices. The CT scanner came out of a project he had undertaken to investigate the practical applications of the mathematical theory of image reconstruction, which dates back to 1917 and also has applications in crystallography, microscopy and radio astronomy.

Initially, Hounsfield conceived of the scanner as a tool of mass screening to detect cancer in the same way that mass miniature X-rays had been used from the 1940s for detecting tuberculosis. However, officials at the Department of Health suggested that the brain might be a better subject; it had the advantage that, unlike the abdomen for example, it could be immobilized for the several minutes that the early scans took. In addition, the brain is difficult to photograph with conventional X-rays because it is encased in the bone of the skull and higher exposure to X-rays is necessary than with other parts of the body; also, except by introducing air or some other contrast medium into the brain it is difficult to tell apart the different structures that make it up.

Hounsfield developed the hardware and software

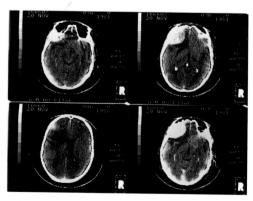

Brain scans taken using the CT scanner
(Atkinson Morley's Hospital)

to convert the 160 readings taken from each of 180 different angles around the head (a total of 28,800) into an image that could be reconstructed on a monitor. One key innovation was to store the readings in a memory device from which the computer would later reconstruct the image in a process that originally took several hours. The other essential development was to write a computer program which could produce a medically useful image using the smallest number of calculations. Working from a specialist knowledge of computer technology on an abstruse area of mathematical theory, Hounsfield was able to produce an innovative piece of medical technology from outside the established community of medical technology.

Although EMI is best known for its entertainments interests (such as its recording of the Beatles, among others) its management has, since the Second World War, pursued a policy of diversification into both civilian and military electronics projects. Their first venture into medical equipment, the thermographic imager, had not been a notable commercial triumph, but the medical equipment market was one in which EMI were keen to become involved. When it became clear both that the CT scanner worked, and that clinicians were enthused by its diagnostic potential, EMI management made a firm decision to enter the medical market, and even acquired two other electro-medical companies to consolidate their position. It has been said that without profits made from the extraordinary expansion of pop music in the 1960s, EMI would never have been able to contemplate going into mass production of the scanner. However, as established medical equipment companies, particularly in Japan, began to supply CT scanners in substantial numbers, EMI began to find the competition difficult. This coincided with a downturn in the popular music market and EMI was forced out of the medical equipment business, selling out to General Electric of the United States in 1980, only eight years after the scanner was first made public.   TMB

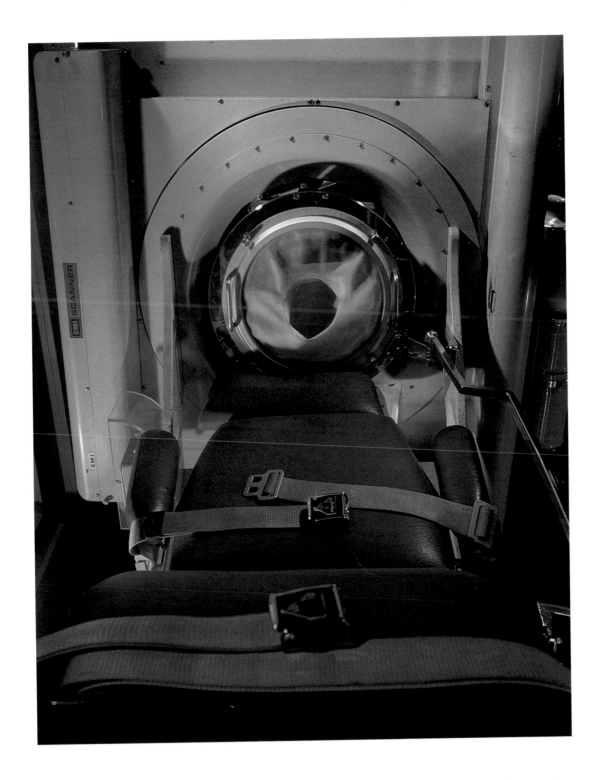

DAVID HOCKNEY: PHOTO-COLLAGE

David Hockney (b. 1937) has had a lifelong love-hate relationship with photography. His father was a keen amateur, and young Hockney first used a box camera on family holidays. As a student and young artist, he used photographs as reference material for his paintings – both those he found in magazines and, increasingly, those he took himself. For Hockney, such photographs were a means to his artistic ends rather than art in themselves; the camera was a convenient recording device. Painting and drawing, where the work of the artist's hand can be clearly traced, were what he really loved.

That attitude began to change in 1969 when he and a friend, Peter Schlesinger, photographed each other sitting at opposite ends of a Paris park bench. Later, Hockney stuck the two snapshots together so that the two men appeared together in the final picture. He had made his first 'joiner' (as he was to call his later, more complicated, collages) and, with very simple means, achieved three complex results.

First, he had made two people sit together when they had not actually done so. Second, by photographing from two slightly different spots, he had shifted his viewpoint and distorted the perspective, yet in a way which is not at all disturbing to the viewer. Finally, and most tellingly, he had suggested something in purely visual terms about the vulnerable, risky nature of a relationship between people who are at once together and apart.

Hockney perceived that 'some of the photographs intended to become paintings didn't have to become paintings', and embarked on a series of 'joiners', rectangular collages of small Polaroid prints. Containing dozens, even hundreds, of Polaroids – painstakingly shot from different viewpoints, closing in on some details, stepping back for others – they were carefully assembled into rectangular grids of white-bordered, square prints. The results were complicated, fascinating and fragmented.

When Hockney progressed to conventional snapshot cameras, the Polaroids looked almost like five-finger exercises, preparations for the real thing. Now, he was able to work more quickly, and to break out of the conventional rectangular picture frame, which he had always found inhibiting. The joiners became increasingly complex in construction, scope, ambition and, most importantly, in their use of time. Prowling restlessly about the people or places which were his subjects, the artist looked at them over an extended period. Nothing could be further from our idea of a photograph as the work of a split-second; in a Hockney joiner, nothing and nobody is quite the same when the picture is finished as it was when it began.

In June 1985, Hockney accepted an invitation from the National Museum of Photography, Film & Television in Bradford to make a joiner of some subject in or around the Museum. He chose the building itself, seen from the same viewpoint as the rather romantic picture of it used by the City of Bradford in its publicity.

The whole exercise took three days, with the actual photography being done, inevitably, in public (several members of the Museum staff and the press appear in the final picture). Hockney and his assistant moved into the Museum to assemble the prints as they were returned by C H Woods, the long-established Bradford photographers and processors. At each session, twenty-five visitors were invited to watch the artist at work, and to share his problems. It was an experience they will not easily forget.

The picture grew more and more complex and Hockney was still taking photographs three days after he began. By then, Woods had closed for the weekend and the final negatives had to be sent to another laboratory. The fact that the colours in these prints were far bluer than in the previous ones did not put Hockney off. Placed on the fringes of the picture, they gave it some of the colours of his California home, and added to the richness of the final result. Like any complex work of art, this piece demands that the viewer take time to look at it.

Hockney is always fascinated by new technologies and, when fax machines and colour laser photocopiers came on to the market, he added them to his armoury, using them in novel and personal ways. In 1990, he acquired a still video camera which he brought to the Museum in 1991, when he exhibited his electronic still portraits, and gave master-classes and a lecture about them.

The electronic stills take Hockney's attempts to overcome what he sees as two major drawbacks of photography even further. They avoid that one-eyed view of the world usually imposed by a camera lens, as his multi-perspective journey down each person's body begins to approach what one might see with one's own eyes. And bringing the camera so close eliminates much of the air between lens and subject. After these direct confrontations, ordinary colour photographs look plastic, dull-toned, and too blue.

Hockney's work is always colourful and inventive, and often witty. His own words describe his aims better than anyone else's. They might also be an appropriate motto for the National Museum of Photography, Film & Television itself:

I do want to make a picture that has meaning for a lot of people. I think the idea of making pictures for twenty-five people in the art world is crazy and ridiculous. It should be stopped . . .

CJF

Opposite: Photo-collage,
Bradford, Yorkshire July *18, 19, 20 1985*

GENETICALLY ENGINEERED MICE

The Science Museum collects artefacts, not organisms. This rule has applied in the Museum since its foundation. But in 1989 the rule was apparently broken when two mice were acquired for its permanent collections. The mice were male, had been preserved by freeze-drying, and were the gift of the Harvard Medical School.

The interest in these mice reflects the revolution in the biological sciences that has accompanied the development of what is often termed genetic engineering. Following Francis Crick and James Watson's discovery of the structure of the genetic material deoxyribose nucleic acid (DNA) in 1953 (*see* CRICK AND WATSON'S DNA MODEL), molecular geneticists made astonishingly rapid progress in unravelling the mechanism of inheritance. By the late 1960s, they knew a great deal about how genetic information is passed down the generations, and they had worked out how this information is translated into specific proteins. In the early 1970s, they acquired the ability to perform a sort of molecular surgery with DNA. Known as recombinant DNA technology, it involves extracting DNA fragments from one organism and inserting them into the genetic structure of another. Over the past fifteen years, recombinant DNA technology has transformed molecular genetics from an elegant but essentially pure science

into a powerful applied science that sits at the heart of biotechnology and molecular medicine. Current applications include the creation of new crop varieties, the development of new drugs, and the production of hormones such as human insulin on an industrial scale.

One particularly important branch of recombinant DNA technology involves the creation of so-called transgenic animals. A transgenic animal is produced when donor DNA fragments are inserted into a newly fertilized egg cell and become incorporated into the animal's genes. As it develops, the inserted DNA is copied along with the animal's own genes, and the result is a transgenic animal with new characteristics. In 1984, Timothy Stewart, Paul Pattengale and Philip Leder of the Harvard Medical School Department of Genetics announced the creation of thirteen strains of transgenic mice. These mice had been modified by the insertion of part of a cancer-causing gene, or oncogene. As a result, the mice were predisposed to develop specific cancers. By working with an 'oncomouse', medical researchers can explore the underlying mechanism of cancer and test potential anti-cancer drugs.

The President and Fellows of Harvard College applied for a patent on Stewart and his colleagues' oncomouse in 1984. A precedent for this patent

application had been set in 1980, when the US Supreme Court decreed that 'anything under the sun that is made by man' is patentable. In 1987 the US Patent Office announced that 'higher life forms' were patentable, and in 1988 it finally granted Harvard's oncomouse patent application. In 1991 the European Patent Office followed its American counterpart by granting a European patent on the Harvard oncomouse. Today the oncomouse is marketed by the Du Pont chemical company for around $100, as compared with around $1 for an ordinary laboratory mouse. The oncomouse is increasingly widely used in cancer research, and other mouse varieties have been developed as models for studying heart disease, AIDS and genetically inherited diseases.

The patenting of life forms is controversial. Some animal welfare and environmental organizations oppose plant and animal patents on ethical grounds; some farming organizations oppose them on commercial grounds. Notwithstanding such objections, the rapid development and commercial application of genetic biotechnologies has provided great scope for plant and animal patents. The Harvard oncomice thus represent an important phase in the development of molecular genetics. With the advent of biotechnology and transgenic animals, it seems that organisms can also be artefacts. JRD

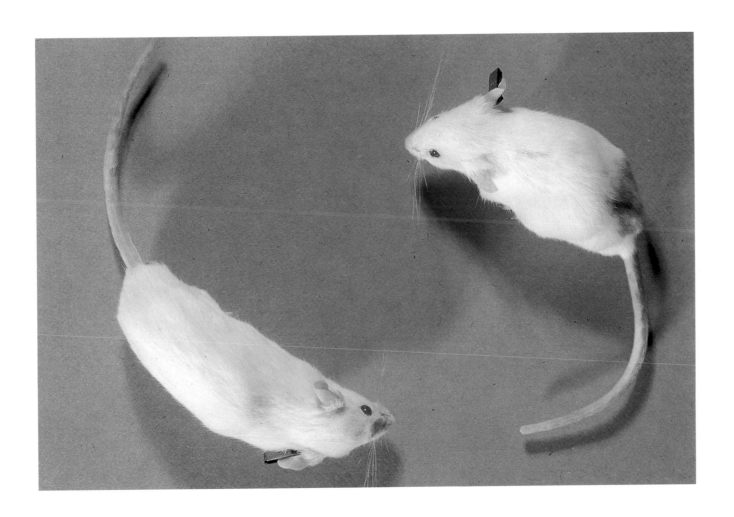

LIST OF CONTRIBUTORS

MA MARCUS AUSTIN
Research Assistant, Special Projects Group

JAB JOHN BAGLEY
Former Senior Curator, Aeronautics

PJB PETER BAILES
Research Assistant, Special Projects Group

EJSB JOHN BECKLAKE
Head of Technology Group

TMB TIM BOON
Curator, Public Health

BPB BRIAN BOWERS
Senior Curator, Electrical Engineering

RB ROGER BRIDGMAN
Curator, Communications

CNB NEIL BROWN
Senior Curator, Classical Physics

RFB ROBERT BUD
Head of Life and Environmental Sciences Group

JB JANE BYWATERS
Manager, Interpretation Unit

SJC SUE CACKETT
Curator, Materials Science

JRC JANET CARDING
Research Assistant, Special Projects Group

JAC JOHN COILEY
Former Head of the National Railway Museum, York

NC NEIL COSSONS
Director, National Museum of Science & Industry

PC PETER CRAIG
Curator, Concorde Exhibition, Fleet Air Arm Museum, Yeovilton

JD JOHN DARIUS
Former Senior Curator, Astronomy & Mathematics

ED ERYL DAVIES
Research Assistant, Special Projects Group

JRD JOHN DURANT
Assistant Director, Science Communication Division

SRE STUART EMMENS
Collections Assistant, Public Health

GPF GRAHAM FARMELO
Head of Education and Interpretation

CJF COLIN FORD
Head of the National Museum of Photography, Film & Television, Bradford

GF GRAEME FYFFE
Collections Development Librarian, Science Museum Library

JG JOHN GRIFFITHS
Head of Special Projects Group

CWH COLIN HARDING
Kodak Curator, National Museum of Photography, Film & Television, Bradford

AJH ALEX HAYWARD
Research Assistant, Special Projects Group

CJH CHRISTINE HEAP
Curator, Information & Support Services, National Railway Museum, York

DWH DIETER HOPKIN
Curator, Collections, National Railway Museum, York

JEI JANE INSLEY
Curator, Environmental Sciences

KLJ KEVIN JOHNSON
Collections Assistant, Astronomy & Mathematics

SAJ STEPHEN JOHNSTON
Research Assistant, Special Projects Group

JK JANE KIRK
Registrar

GML GHISLAINE LAWRENCE
Senior Curator, Clinical Medicine

JL JOHN LIFFEN
Collections Assistant, Road and Rail Transport

PRM PETER MANN
Senior Curator, Road and Rail Transport

AQM ALAN MORTON
Senior Curator, Modern Physics

SM SUE MOSSMAN
Research Assistant, Special Projects Group

GM GRAHAM MOTTRAM
Deputy Director and Curator, Fleet Air Arm Museum, Yeovilton

ALN ANDREW NAHUM
Senior Curator, Aeronautics

AKN ANN NEWMARK
Senior Curator, Documentation

FR FRANCESCA RICCINI
Research Assistant, Special Projects Group

PR PIPPA RICHARDSON
Interpretation Officer

DAR DEREK ROBINSON
Head of Science Group

JCR JOHN ROBINSON
Senior Curator, Water Transport

AES TONY SALE
Manager, Computer Conservation Programme

GRS ROSS SHARP
Aviation Operations Officer, Science Museum, Wroughton

WS WENDY SHERIDAN
Curator, Pictorial Collection

RSB ROB SHORLAND-BALL
Deputy Head, National Railway Museum, York

PDS PETER STEPHENS
Former Curator, Civil Engineering

DDS DORON SWADE
Senior Curator, Computing

RT ROGER TAYLOR
Senior Curator, Photography, National Museum of Photography, Film & Television, Bradford

JT JOHN TRENOUTH
Senior Curator, Television, National Museum of Photography, Film & Television, Bradford

PJT PETER TURVEY
Research Assistant, Special Projects Group

RV ROD VARLEY
Senior Curator, Film, National Museum of Photography, Film & Television, Bradford

DV DENYS VAUGHAN
Senior Curator, Time Measurement

JAW JANE WESS
Curator, Astronomy & Mathematics

AWW ANTHONY WILSON
Publications Manager

DJW DAVID WOODCOCK
Collections Assistant, Electrical Engineering

MTW MICHAEL WRIGHT
Curator, Mechanical Engineering

TW TOM WRIGHT
Assistant Director, Collections Division

The main images of Science Museum objects accompanying the articles were specially commissioned from the photographer Philip Sayer.

BIBLIOGRAPHY

BYZANTINE SUNDIAL-CALENDAR

J V Field, 'Some Roman and Byzantine portable sundials and the London sundial-calendar', *History of Technology*, Vol 12, 1990, pp 103–35

J V Field & M T Wright, 'Gears from the Byzantines', *Annals of Science*, Vol 42, 1985, pp 87–138

D R Hill, 'Al-Bīrūnī's Mechanical Calendar', *Annals of Science*, Vol 42, 1985, pp 139–63

D J de S Price, 'Gears from the Greeks', *Transactions of the American Philosophical Society*, New Series, Vol 64, Pt. 7, 1974

M T Wright, 'Rational and irrational reconstruction: the London sundial-calendar and the early history of geared mechanisms', *History of Technology*, Vol 12, 1990, pp 65–102

ISLAMIC GLASS ALEMBIC

R G W Anderson, 'Early Islamic chemical glass', *Chemistry in Britain*, 1983, p 822

THE GIUSTINIANI MEDICINE CHEST

John Burnett, 'The Giustiniani Medicine Chest', *Medical History*, Vol 26, 1982, pp 325–33

THE STANDARDS OF THE REALM

R D Connor, *The Weights and Measures of England*, HMSO, 1987

F G Skinner, *Weights and Measures*, HMSO, 1967

R E Zupko, *British Weights and Measures*, University of Wisconsin Press, 1977

SLIDE RULE BY ROBERT BISSAKER

Anthony Turner, *Early Scientific Instruments. Europe 1400–1800*, London, 1987, pp 161–5

NAPIER'S BONES

E M Horsburgh (ed), *Handbook of the Exhibition of Napier Relics and of Books, Instruments and Devices for Facilitating Calculation*, Royal Society of Edinburgh, 1914. Reprinted as *Handbook of the Napier Tercentenary Celebration of Modern Instruments and Methods of Calculation*, Charles Babbage Institute Reprint Series for the History of Computing, Vol 3, Tomash Publishers, 1982

Michael R Williams, *A History of Computing Technology*, Prentice-Hall, 1985, Chapter 2

HAUKSBEE'S AIR PUMP

F Hauksbee, *Physico-Mechanical Experiments on Various Subjects*, 1709, (2nd ed. Johnson Reprint Corporation, New York and London, 1970)

THE ORIGINAL ORRERY

H C King and J R Millburn, *Geared to the Stars*, University of Toronto Press, 1978

SISSON'S RULE

Major General Roy, 'Account of the Measurement of a Base on Hounslow-Heath', *Philosophical Transactions*, Vol 75, 1785, p 401

SHELTON'S ASTRONOMICAL REGULATOR

D Howse, 'Captain Cook's pendulum clocks', *Antiquarian Horology*, Vol 6, 1969, pp 62–76

D Howse and B Hutchinson, 'The saga of the Shelton clocks', *Antiquarian Horology*, Vol 6, 1969, pp 281–98

H Woolf, *The Transits of Venus*, Princeton University Press, 1959

ARKWRIGHT'S SPINNING MACHINE

W English, *The Textile Industry*, Longman Green & Co, 1969

J Tann, *The Development of the Factory*, London, 1970

TROUGHTON'S DIVIDING ENGINE

David Brewster (ed), 'Graduation', *The Edinburgh Encyclopedia*, Vol 10, Edinburgh, 1830

Allan Chapman, *Dividing the Circle*, Ellis Horwood, 1990

Jesse Ramsden, *Description of an Engine for Dividing Mathematical Instruments*, London, 1777

L T C Rolt, *Tools for the Job*, HMSO, 1986

HERSCHEL'S SEVEN-FOOT TELESCOPE

Mrs J Herschel, *Memoirs and Correspondences of Caroline Herschel*, John Murray, 1876

H C King, *The History of the Telescope*, Dover Publications Inc, 1979

M B Ogilvie, *Women in Science*, MIT Press, 1986

BOULTON AND WATT ROTATIVE ENGINE

H W Dickinson and R Jenkins, *James Watt and the Steam Engine*, Oxford University Press, 1927

J Farey, *The Steam Engine*, London, 1827

R J Law, *The Steam Engine*, HMSO, 1965

R J Law, *James Watt and the separate condenser*, HMSO, 1976

J Tann, 'Boulton and Watt's organisation of steam engine production before the opening of the Soho foundry', *Transactions of the Newcomen Society*, Vol 49, 1978

SYMINGTON'S MARINE ENGINE

W S Harvey, *William Symington, Inventor and Engine Builder*, Northgate, 1980

RAMSDEN'S THREE-FOOT THEODOLITE

Account of the Observations and Calculations of the Principal Triangulation, Ordnance Survey, 1858

MAUDSLAY'S SCREW-CUTTING LATHE

K R Gilbert, *Henry Maudslay: Machine Builder*, HMSO, 1971

C Holtzapffel, *Turning and Mechanical Manipulation*, Vol 2, London, 1846

S Smiles, *Industrial Biography: Iron Workers and Tool Makers*, John Murray, 1863

HERSCHEL'S PRISM AND MIRROR

E S Cornell, 'The radiant heat spectrum from Herschel to Melloni – I Herschel and his contemporaries', *Annals of Science*, Vol 3, 1938, pp 119–37

William Herschel, 'Investigation of the powers of the prismatic colours to heat and illuminate objects', *Philosophical Transactions of the Royal Society*, 1800, Pt 2, pp 255–83

William Herschel, 'Experiments on the refrangibility of the invisible rays of the Sun' *Philosophical Transactions of the Royal Society*, 1800, Pt 2, pp 284–93

William Herschel, 'Experiments on the solar and on the terrestrial rays that occasion heat, Pt 1', *Philosophical Transactions of the Royal Society*, 1800, Pt 2, pp 293–326

William Herschel, 'Experiments on the solar and on the terrestrial rays that occasion heat, Pt 2', *Philosophical Transactions of the Royal Society*, 1800, Pt 3, pp 437–538

'COALBROOKDALE BY NIGHT'

Ralph Hyde, *Panoramania!*, Trefoil for the Barbican Art Gallery, 1988

Francis D Klingender (edited and revised by Arthur Elton), *Art and the Industrial Revolution*, Evelyn, Adams & Mackay, 1968

Stuart B Smith, *A View from the Iron Bridge*, Ironbridge Gorge Museum Trust, 1979

PORTSMOUTH BLOCK-MAKING MACHINERY

R Beamish, *Memoir of the Life of Sir Marc Isambard Brunel*, 2nd edition, Longman, 1862

C C Cooper, 'The production line at Portsmouth block mill', *Industrial Archaeology Review*, Vol 6, No 1, 1982, pp 28–44

K R Gilbert, *The Portsmouth Block-Making Machinery*, Science Museum Monograph, 1965

Abraham Rees (ed), 'Machinery for manufacturing ships' blocks', *The Cyclopædia*, 1819

TREVITHICK'S HIGH-PRESSURE ENGINE

H W Dickinson, *A Short History of the Steam Engine*, 1938

John Farey, *A Treatise on the Steam Engine*, Vol 2, David & Charles, 1971

L Ince, 'Richard Trevithick's patent steam engine', *Stationary Power*, Vol 1, 1984

Abraham Rees (ed), 'Steam-engine', *The Cyclopædia*, 1819

THE 'COMET' STEAM ENGINE

Martin Hughson, *John Robertson, Engineer*, Bearhead & Neilston Historical Association, 1989

'PUFFING BILLY'

Dendy Marshall, *Two Essays in Early Locomotive History*, The Locomotive Publishing Co Ltd, 1928

DAVY'S SAFETY LAMP

Sir Humphry Davy, *On the Safety Lamp for Coal Miners with Some Researches on Flame*, London, 1818

F W Hardwick and L T O'Shea, 'Notes on the history of the safety lamp', *Transactions of the Institution of Mining Engineers*, Vol 51, Pts 5 & 6, 1916, pp 548–724

LISTER'S 1826 MICROSCOPE

B Bracegirdle, 'Famous Microscopists: Joseph Jackson Lister, 1786–1869', *Proceedings of the Royal Microscopical Society*, Vol 22, September 1987, pp 273–97

R B Fisher, *Joseph Lister 1827–1912*, Macdonald & James, 1977

BELL'S REAPER

The Revd P Bell, 'Some Account of "Bell's Reaping Machine"', *Quarterly Journal of Agriculture*, 1854, pp 185–95

G E Fussell, *The Farmer's Tools*, 1952

L J Jones, 'The Early History of Mechanical Harvesting', *History of Technology*, Vol 4, 1979, pp 101–48

STEPHENSON'S 'ROCKET'

B Reed, *Locomotives in Profile*, Vol 1, Profile Publications, 1971

R G H Thomas, *The Liverpool & Manchester Railway*, Batsford, 1980

BABBAGE'S CALCULATING ENGINES

Allan G Bromley, 'Difference engines and analytical engines', *Computing Before Computers*, (ed William Aspray), Iowa State University Press, 1990

Doron Swade, *Charles Babbage and his Calculating Engines*, Science Museum, 1991

TALBOT'S 'LATTICED WINDOW'

H J P Arnold, *William Henry Fox Talbot*, London, 1977

COOKE AND WHEATSTONE'S TELEGRAPH

Brian Bowers, *Sir Charles Wheatstone*, HMSO, 1975

Geoffrey Hubbard, *Cooke and Wheatstone and the Invention of the Electric Telegraph*, Routledge & Kegan Paul, 1965

THE BROUGHAM

G N Hooper, 'Transition in London carriages', paper read at the Institute of British Carriage Manufacturers, 23 January 1896

EARLY DAGUERREOTYPES OF ITALY

Helmut Gernsheim, *The Origins of Photography*, London & New York, 1982

THE ROSSE MIRROR

H C King, *The History of the Telescope*, Dover Publications Inc, 1979

The Illustrated London News, 9 Sept 1843, pp 165–6

The Illustrated London News, 19 April 1845, pp 253–4

Earl of Rosse, 'Observations on the nebulae', *Philosophical Transactions of the Royal Society*, 1850 pp 499–514

ELIAS HOWE'S SEWING MACHINE

Sarah Levitt, *Victorians Unbuttoned*, George Allen & Unwin, 1986

Grace Rogers Cooper, 'The invention of the sewing machine', *United States National Museum Bulletin*, No 254, Smithsonian Institution, 1968

JOULE'S PADDLE-WHEEL APPARATUS

D S L Cardwell, *From Watt to Clausius*, London, 1971

D S L Cardwell, *James Joule: A Biography*, Manchester and New York, 1989

J P Joule, *Scientific Papers*, London, 1884

H J Steffens, *James Prescott Joule and the Concept of Energy*, New York, 1979

ORIGINAL MAUVE DYE

E Farber (ed), *Great Chemists*, Interscience, 1961, pp 758–72

C C Gillispie (ed), *Dictionary of Scientific Biography*, Vol 10, Charles Scribner's Sons, 1974, pp 515–17

D H Leaback, 'Perkin's pioneering enterprise', *Chemistry in Britain*, 1988, pp 787–90

Obituary Notices 'William Henry Perkin', *Journal of the Chemistry Society*, Vol 93 (Pt II), 1908, pp 2214–57

A S Travis, 'Perkin's mauve: ancestor of the organic chemical industry', *Technology & Culture*, Vol 31, 1990, pp 51–82

LEWIS CARROLL'S PHOTOGRAPHS

'Lewis Carroll Photographer', National Museum of Photography, Film & Television, 1987

THOMSON'S MIRROR GALVANOMETER

Bernard S Finn, *Submarine Telegraphy, The Grand Victorian Technology*, HMSO, 1973

THE KEW PHOTOHELIOGRAPH

W De La Rue, 'On the total solar eclipse of July 18th, 1860 observed at Rivabellosa, near Miranda de Ebro, in Spain [Bakerian Lecture]', *Philosophical Transactions of the Royal Society*, Vol 52, 1862, pp 333–416

W De La Rue, 'Comparison of Mr De La Rue's and Padre Secchi's eclipse photographs', *Proceedings of the Royal Society*, Vol 13, 1864, pp 442–4

H C King, *The History of the Telescope*, Dover Publications Inc, 1979

THE FIRST PLASTIC

R D Friedel, *Men, Materials, and Ideas: A History of Celluloid*, The Johns Hopkins University, PhD, 1977, Xerox University Microfilms, Ann Arbor, Michigan, 1978

S Katz, *Classic Plastics*, Thames & Hudson, 1984

M Kaufman, *The First Century of Plastics*, Plastics Institute, London, 1960

THE LENOIR GAS ENGINE

C L Cummins, *Internal Fire: The Internal Combustion Engine 1673–1900*, Oregon, 1976

J Day, *Engines: The Search for Power*, London, 1980

W Robinson, *Gas and Petroleum Engines*, 2nd ed. London, 1902

BESSEMER CONVERTER

K C Barraclough, *Steelmaking, 1850–1900*, Institute of Metals, 1990

H Bessemer, *An Autobiography*, Facsimile edition, Institute of Metals, 1989

W K V Gale, *Iron and Steel*, Longman, 1969

'SWAN' AND 'RAVEN'

T Wright, 'Scale models, similitude, and dimensions: aspects of mid-nineteenth-century engineering science', *Annals of Science*, Vol 49, 1992, pp 233–54

HOLMES' LIGHTHOUSE GENERATOR

J M Douglass, 'The Electric Light Applied to Lighthouse Illumination', *Minutes of the Proceedings of the Institution of Civil Engineers*, Vol 57, Pt III, 1878–9

W J King, 'The development of electrical technology in the 19th Century: 3. The early arc light and generator', *Contributions from the Museum of History and Technology*, United States National Museum Bulletin, No 228, paper 30, Smithsonian Institution, 1962, pp 333–407

JULIA MARGARET CAMERON'S 'IAGO'

Colin Ford, *The Cameron Collection*, Van Nostrand Reinhold/National Portrait Gallery, 1975

CROOKES' RADIOMETER

W Crookes, 'On the illumination of lines of molecular pressure and the trajectory of molecules', *Philosophical Transactions of the Royal Society*, Vol 170, 1879, pp 135–64

W Crookes, 'Molecular physics in high vacua', *Proceedings of the Royal Institution of Great Britain*, Vol 9, 1879–81, pp 138–59

W Crookes, 'On radiant matter', *Nature*, Vol 20, 1879, pp 419–23, 436–40

E E Fournier D'Albe, *The Life of Sir William Crookes*, London, 1923

BELL'S OSBORNE TELEPHONE

Robert V Bruce, *Bell*, Gollancz, 1973

Ithiel de Sola Pool (ed), *The social impact of the telephone*, MIT Press, 1977

WIMSHURST'S ELECTROSTATIC MACHINE

Alfred W Marshall, *The Wimshurst Machine: how to make it and use it: a practical handbook on the construction and working of the Wimshurst machine*, Lindsay Publications, 1989, 2nd edition (facsimile reprint of 1908 original)

Margaret Rowbottom and Charles Susskind, *Electricity and Medicine: History of their Interaction*, Macmillan, 1984

THE FIRST TURBO-GENERATOR

B P Bowers, *History of Electric Light and Power*, Peter Peregrinus/Science Museum, 1982

W Garret Scaife, 'The Parsons steam turbine', *Scientific American*, Vol 252, No 4, April 1985

ROVER SAFETY BICYCLE

H W Bartleet, *Bartleet's Bicycle Book*, London, 1931, reprinted 1983 by John Pinkerton

C F Caunter, *The History and Development of Cycles. Part I*, HMSO, 1955

C F Caunter, *Handbook of the Collection Illustrating Cycles. Part II*, HMSO, 1958

Cyclist and Wheel World Annual, 1862

A Sharp, *Bicycles & Tricycles – An Elementary Treatise on their Design and Construction*, MIT Press, 1977, reprint of Longmans 1896 edition

J K Starley, 'The evolution of the cycle', *Journal, Society of Arts*, Vol 46, No 2374, 20 May 1898

H Sturmey, *Sturmey's Indispensable Handbook to the Safety Bicycle, Treating of Safety Bicycles, Their Varieties, Construction & Use*, Iliffe & Son, 1885, reprinted 1982 by John Pinkerton

THE KODAK CAMERA

Brian Coe, *Kodak Cameras – The First Hundred Years*, Hove Photo Books, 1988

Colin Ford (ed), *The Kodak Museum: The Story of Popular Photography*, Century, 1989

EARLY CINE-CAMERAS

Christopher Rawlence, *The Missing Reel: The Untold Story of the Lost Inventor of Moving Pictures*, Collins, 1990

LOCOMOTIVE FROM FIRST TUBE RAILWAY

P P Holman, *The Amazing Electric Tube*, London Transport Museum, 1990

M A C Horne, *The Northern Line: A Short History*, Douglas Rose/Nebulous Books, 1987

A A Jackson and D F Croome, *Rails Through the Clay*, George Allen & Unwin Ltd, 1962

T S Lascelles, *The City & South London Railway*, The Oakwood Press, 1955

PARSONS MARINE STEAM TURBINE

A Richardson, *The Evolution of Parsons Steam Turbine*, London, 1911

PANHARD ET LEVASSOR CAR

The Saturday Review, 20 July 1895, pp 79–80

'COMING SOUTH, PERTH STATION'

C Hamilton Ellis, *Railway Art*, Ash & Grant, 1977

Aubrey Noakes, *William Frith: Extraordinary Victorian Painter*, Jupiter, 1978

Jeffrey Richards and John Mackenzie, *The Railway Station – A Social History*, Oxford University Press, 1986

Train Spotting – Images of the Railway in Art, Nottingham Castle Museum, 1985

THOMSON'S CATHODE RAY TUBE

A J G Hey and P Walters, *The Quantum Universe*, Cambridge University Press, 1987

A Pais, *Inward Bound*, Oxford University Press, 1986

MARCONI'S FIRST TUNED TRANSMITTER

W J Baker, *A History of the Marconi Company*, Methuen, 1970

Sir Eric Eastwood (ed), *Wireless Telegraphy*, (Royal Institution Library of Science), Applied Science Publishers, 1974

POULSEN'S TELEGRAPHONE

Marvin Camras (ed), *Magnetic Tape Recording*, Van Nostrand Rheinhold, 1985

'The Telegraphone', *The Electrician*, 26 April 1901, pp 5–7 and 31 July 1903, pp 611–12

FLEMING'S ORIGINAL THERMIONIC VALVES

J A Fleming, 'A further examination of the Edison effect in glow lamps', *Philosophical Magazine*, 5th series, Vol 42, No 254, 1896

HABER'S SYNTHETIC AMMONIA

L F Haber, *The Chemical Industry 1900–1930. International Growth and Technological Change*, The Clarendon Press, 1971

'THE MUNITION GIRLS'

Henry Baker, *The 'Steel' Bakers of Rotherham: being a brief history of the Baker family and the business of John Baker & Bessemer Ltd*, Rotherham, 1960

R T Wilson and E J Twigg, *Industrial Sheffield and Rotherham: The Official Handbook of the Sheffield and Rotherham Chambers of Commerce*, Derby, 1919

ASTON'S MASS SPECTROGRAPH

F W Aston, 'A simple form of micro-balance for determining the densities of small quantities of gases', *Proceedings of the Royal Society of London*, Series A, Vol 89, 1914, pp 439–46

C C Gillispie (ed), *Dictionary of Scientific Biography*, Vol 1, Charles Scribner's Sons, 1970, pp 320–2

F N Magill (ed), *The Nobel Prize Winners: Chemistry*, Salem Press, 1990, pp 250–7

Obituary Notices of Fellows of the Royal Society, Vol 5, pp 634–50

ALCOCK AND BROWN'S VICKERS VIMY

Sir John Alcock and Sir Arthur Whitten Brown, *Our Transatlantic Flight*, Kimber, 1969

AUSTIN SEVEN PROTOTYPE CAR

R J Wyatt, *The Austin Seven*, David & Charles, 1968

LOGIE BAIRD'S TELEVISION APPARATUS

R W Burns, *British Television, The Formative Years*, London, 1986

G R M Garratt, *Television*, Science Museum, 1937

SUPERMARINE S.6B FLOATPLANE

Flight, 1931

Jane's All the World's Aircraft, 1931

Wing-Commander A H Orlebar, *Schneider Trophy*, London, 1933

LAWRENCE'S ELEVEN-INCH CYCLOTRON

F Close, M Marten and C Sutton, *The Particle Explosion*, Oxford University Press, 1987

R Rhodes, *The Making of the Atomic Bomb*, Simon and Schuster, 1986

COCKCROFT AND WALTON'S ACCELERATOR

J G Crowther, *The Cavendish Laboratory, 1874–1974*, Macmillan, 1974

CHADWICK'S PARAFFIN WAX

Emilio Segrè, *From X-rays to Quarks, Modern Physicists and Their Discoveries*, W H Freeman, 1980

DISCOVERY OF POLYETHYLENE

Carol Kennedy, *ICI. The Company that Changed Our Lives*, Hutchinson, 1986

THE BOEING 247D

Peter Berry, *The Douglas Commercial Story*, Air-Britain Publications, 1971

John W R Taylor and Kenneth Munson, *Air Transport Before the Second World War*, New English Library, 1975

DISCOVERY OF ARTIFICIAL RADIOACTIVITY

Spencer R Weart, *Scientists in Power*, Harvard University Press, 1979

REYNOLDS' CINERADIOGRAPHY APPARATUS

Russell J Reynolds, 'Cineradiography', *Institution of Electrical Engineers Journal*, Vol 79, No 478, October 1936, pp 389–400

ORIGINAL RADAR RECEIVER

E G Bowen, *Radar Days*, Adam Hilger, 1987

Russell Burns (ed), *Radar Development to 1945*, Peter Peregrinus, 1988

Sir Robert Watson-Watt, *Three Steps to Victory*, London, 1957

MANCHESTER DIFFERENTIAL ANALYSER

S Auergarten, *Bit by Bit – An Illustrated History of Computers*, Unwin, 1984

V Bush, *Pieces of the Action*, Morrow, 1970, p 161

J Crank, *The Differential Analyser*, Longmans, 1947

D Hartree, 'The differential analyser', *Nature*, Vol 135, June 1935, pp 940–3

D Hartree, 'A great calculating machine', *Proceedings of the Royal Institution of Great Britain*, Vol 31, 1939–40, p 151

'Machine solves mathematical problems', *Meccano Magazine*, No 19, June 1934, pp 442–4

EMITRON CAMERA TUBE

R W Burns, *British Television – The Formative Years*, 1986

Bruce Norman, *Here's Looking At You*, Royal Television Society/BBC, 1984

E Pawley, *BBC Engineering 1922–1972*, BBC, 1972

'MALLARD'

C J Allen, *The Gresley Pacifics of the LNER*, Ian Allan, 1950

J Bellwood and D Jenkinson, *Gresley and Stanler*, HMSO, 1986, 2nd edition

O S Nock, *The Gresley Pacifics*, David & Charles, 1982

M Rutherford, *Mallard the Record Breaker*, Newburn House and the Friends of the National Railway Museum, 1988

RANDALL AND BOOT'S CAVITY MAGNETRON

P M Rolph (ed), *Fifty Years of the Cavity Magnetron*, School of Physics and Space Research, University of Birmingham, 1991

Robert L Wathen, 'Genesis of a generator – the early history of the magnetron', *Journal of the Franklin Institute*, Vol 255, No 4, April 1953

THE ROLLS-ROYCE MERLIN

Alec Harvey-Bailey, *The Merlin in Perspective – the Combat Years*, Rolls-Royce Heritage Trust, 1983

L J K Setright, *The Power to Fly*, Allen and Unwin, 1971

GLOSTER-WHITTLE E.28/39 JET AIRCRAFT

John Golley, *Whittle – The True Story*, Airlife, 1987

Sir Frank Whittle, *Jet*, Frederick Muller, London, 1953

THE V2 ROCKET

Gregory Kennedy, *Vengeance Weapon 2 – The V2 Guided Missile*, published for the National Air and Space Museum by the Smithsonian Institution Press, 1983

Frederick Ordway III and Mitchell R Sharpe, *The Rocket Team*, William Heinemann, 1979

THE FIRST MARINE GAS TURBINE

K Phelan and M H Brice, *Fast Attack Craft*, Macdonald & James, 1977

PILOT ACE

M Campbell-Kelly, 'Programming the Pilot ACE: Early programming activity at the National Physical Laboratory', *Annals of the History of Computing*, Vol 3, No 2, April 1981, pp 133–62

The Early British Computer Conferences, Vol 14, Charles Babbage Institute reprint series, MIT Press, London, 1989

Michael R Williams, *A History of Computing Technology*, Prentice-Hall, 1985

NMR ELECTROMAGNET

Christine Sutton, 'A magnetic window into bodily functions', *New Scientist*, Vol 3, No 1525, September 1986, pp 32–7

CRICK AND WATSON'S DNA MODEL

Robert Olby, *The Path to the Double Helix*, Macmillan, 1974

J D Watson and F H C Crick, 'The double helix', *Nature*, Vol 171, April 1953, pp 737–8

J D Watson, *The Double Helix*, Weidenfeld & Nicolson, 1968. (Reprinted with additional articles, edited by Gunther S Stent, Weidenfeld & Nicolson, 1981)

FLYING BEDSTEAD TEST RIG

Mike Rogers, *VTOL Military Research Aircraft*, Haynes Foulis Aviation, 1989

THE CAESIUM ATOMIC CLOCK

L Essen and J V L Parry, 'The caesium resonator as a standard of frequency and time', *Philosophical Transactions of the Royal Society*, Series A, Vol 250, 1957, pp 45–69

P Forman, 'Atomichron: the atomic clock from concept to commercial product', *Proceedings of the IEEE*, Vol 73, 1985, pp 1181–204

THE WORLD'S FIRST HOVERCRAFT

P R Crewe, 'The hovercraft – a new concept in maritime history', *Transactions of the Royal Institution of Naval Architects*, Vol 102, 1960, pp 315–65

AMPEX VR1000 VIDEO RECORDER

E Pawley, *BBC Engineering 1922–1972*, BBC, 1972

J Robinson, *Videotape Recording*, Vocal Press, 1975

FERRANTI PEGASUS COMPUTER

Simon Lavington, *Early British Computers*, Manchester University Press, 1980

INTERNAL CARDIAC PACEMAKER

R W Portal, J G Davies, A Leatham, A H M Siddons, 'Artificial pacing for heart block', *The Lancet*, Vol 2, 1962, pp 1369–75

SCANNING ELECTRON MICROSCOPE

C W Oatley, 'The early history of the scanning electron microscope', *Journal of Applied Physics*, Vol 53, 1982, pp R1–13

A D G Stewart, 'The origins and development of scanning electron microscopy', *Journal of Microscopy*, Vol 139, 1985, pp 121–7

ULTRASOUND SCANNER

Stuart S Blume, *Insight & Industry – On the Dynamics of Technological Change in Medicine*, MIT Press, 1992

Ann Oakley, *The Captured Womb – A History of the Medical Care of Pregnant Women*, Oxford, 1984

Edward Yoxen, 'Seeing with sound: a study of the development of medical images', *The Social Construction of Technological Systems*, (ed Wiebe E Bijker et al), MIT Press, 1987

THE FIRST FLEXIBLE MANUFACTURING SYSTEM

D T N Williamson, 'New concepts of manufacture', *The Engineer*, September 1967, p 342

CONCORDE 002

D Edgerton, *England and the Aeroplane*, Macmillan, 1991

C Orlebar, *The Concorde Story*, Temple Press, 1986

K Owen, *Concorde, New Shape in the Sky*, Janes/Science Museum, 1982

ROLLS-ROYCE RB 211

Bill Gunston, *Rolls-Royce Aero Engines*, Patrick Stephens Ltd, 1989

Sir Stanley Hooker, *Not Much of an Engineer*, Airlife, 1984

THE FIRST BRAIN SCANNER

Stuart S Blume, *Insight & Industry – On the Dynamics of Technological Change in Medicine*, MIT Press, 1992

G N Hounsfield, 'Computerized axial scanning (tomography): Part 1. Description of system', *British Journal of Radiology*, Vol 46, 1973, pp 1016–22

Charles Süsskind, 'The Invention of Computed Tomography', *History of Technology*, Vol 6, No 1, 1981, pp 39–80

DAVID HOCKNEY: PHOTO-COLLAGE

David Hockney and Paul Joyce, *Hockney on photography*, Jonathan Cape, 1988

GENETICALLY ENGINEERED MICE

Sheldon Krimsky, *Biotechnics and Society: The Rise of Industrial Genetics*, Praeger, 1991

INDEX